Automatic Control Simulation of Aircraft Traction Cascade System

飞机牵引级联系统自动控制仿真

（英文版）

刘天畅　著

中山大学出版社

·广州·

图书在版编目（CIP）数据

飞机牵引级联系统自动控制仿真 = Automatic Control Simulation of Aircraft Traction Cascade System：英文/刘天畅著 . —广州：中山大学出版社，2022. 12
ISBN 978 - 7 - 306 - 07645 - 8

Ⅰ. ①飞… Ⅱ. ①刘… Ⅲ. ①旅客机—牵引系统—研究—英文 Ⅳ. ①V271. 1

中国版本图书馆 CIP 数据核字（2022）第 215904 号

FEIJI QIANYIN JILIANXITONG ZIDONG KONGZHI FANGZHEN

出 版 人：王天琪
策划编辑：曹丽云　熊锡源
责任编辑：熊锡源
封面设计：曾　婷
责任校对：李昭莹
责任技编：靳晓虹
出版发行：中山大学出版社
电　　话：编辑部 020 - 84111997，84110283，84113349
　　　　　发行部 020 - 84111998，84111981，84111160
地　　址：广州市新港西路 135 号
邮　　编：510275　　　　传　真：020 - 84036565
网　　址：http：//www. zsup. com. cn　E-mail：zdcbs@ mail. sysu. edu. cn
印 刷 者：广东虎彩云印刷有限公司
规　　格：787mm×1092mm　1/16　13 印张　285 千字
版次印次：2022 年 12 月第 1 版　　2022 年 12 月第 1 次印刷
定　　价：55. 00 元

内 容 简 介

在机场建设水平迅速提高和建设规模不断扩大的背景下，本书针对大型客机的研发、试飞和运营的需求，采用机器学习、5G 数据链和智能图像处理等方法研究了飞机牵引级联系统自动控制仿真。通过机器学习建立了牵引和推动两类模型，并设计了基于 5G 传输链路的通讯控制协议；在控制过程的研究中，采用机器学习、智能图像处理、数据集构建和 5G 数据传输等方法，研究了飞机牵引级联系统的稳定性、飞机牵引运动过程的环境识别和运动状态控制；通过仿真实现了飞机牵引的自动控制。

本书可供从事自动控制、模式识别与智能图像处理、5G 数据传输等相关专业的研究人员和工程技术人员参考。

Abstract

With the rapid improvement and the continuous expansion of China's airport construction, many new requirements have been put forward for China's large passenger aircraft development, test flight and operation. This book takes the aircraft tractor automatic control system as the research object, applying machine learning, high-speed data link and intelligent image processing to aircraft tractor automatic control simulation. Machine learning is used to establish the towing and pushing models in order to realize the aircraft tractor automatic control, and the communication control protocols based on the high-speed transmission links are designed. Machine learning, intelligent image processing, data set construction and high-speed data transmission have been adopted to study the stability of the aircraft traction cascade system, the environment recognition of the aircraft tractor motion and the motion state control in the simulation control process. Simulation is used to study the aircraft traction cascade system, and the automatic control of the aircraft traction has been realized.

The book can be used as a reference for researchers and engineers who are engaged in automatic control, pattern recognition and intelligent image processing, 5G high-speed data transmission and other related majors.

前　　言

随着社会、经济和科技的快速发展，机场建设水平不断提高，机场建设规模不断扩大，民航客运运营压力日益增加，对飞机牵引自动控制发展提出了新的需求和挑战。以大型客机的研发、试飞和运营为契机，深入研究智能型飞机牵引车及其自动控制系统，对航空工业的发展具有重要意义。

本书通过飞机牵引级联系统仿真设计，确定了仿真系统功能指标和参数指标；通过仿真实验研究了牵引过程的运动方式，建立了多类输入的神经网络；对运行过程中的图像信息、姿态信息、类别信息进行识别训练，生成推动和牵引模型，用于飞机牵引车的自动控制，并使用特征检测、水雾模型检测、图像分块式检测等方法对飞机牵引车运动过程中的环境信息进行识别，对牵引车的运动状态进行约束；设计了基于 5G 数据链路的传输协议，并对控制数据和视频数据的传输速率和准确率进行了验证，实现了本地数据的云上分析计算，适用于外场试飞任务中的大规模数据卸载、多路高清视频传输和多种控制指令校验。

全书共八章。第一章对飞机牵引级联系统及其涉及的飞机牵引车传动与控制方式、飞机出入库防撞及自动泊位系统、机场 FOD 智能检测进行概述；第二章介绍机器学习的发展历程，综述了飞机牵引级联系统涉及的机器学习模型、数据处理与相关算法；第三章采用仿真方法，设定飞机牵引级联系统功能指标和参数指标；第四章研究飞机牵引级联系统 PID 控制及其稳定性；第五章基于机器学习研究飞机牵引级联系统的运动控制和环境约束控制；第六章研究高速数据传输链路在飞机牵引自动控制中的应用；第七章对飞机牵引级联系统自动控制进行仿真研究；第八章对研究结果进行归纳总结，并展望了飞机牵引车智能化自动控制的未来发展方向。

在研究过程中，笔者得到了上海交通大学周越教授和中国商飞民用飞机试飞中心冯灿高级工程师的帮助，在此深表感谢！感谢上海交通大学电子信息与电气工程学院吴沂军等老师和同学在研究过程中给予的无私帮助！在研究过程中，笔者参考了大量文献资料，在此，向所有文献的作者表示衷心感谢！

希望本书的出版能够促进大型客机飞机牵引车自动控制技术的发展。由于笔者水平有限，书中难免存在疏漏及不足之处，恳请读者批评指正！

刘天畅

2021 年 12 月于中国商飞

Preface

With the rapid development of society, economy, science and technology, the airport construction level is constantly improving, the construction scale is expanding and the operation pressure of civil aviation passenger transport is increasing, which puts forward new demands and challenges for the development of aircraft tractor automatic control. It is of great significance for the development of aviation industry to study the intelligent aircraft tractor and its automatic control system taking the development of large passenger aircraft, test flight and its operation as an opportunity.

In the research, the function and parameter indicators of the simulation system were determined through the simulation design of the aircraft tractor cascade system. The motion mode of traction process was studied by simulation experiments, and a multi-input neural network was established. The image information, attitude information and category information were trained in the operation process, the driving model and traction model were generated for the automatic control of aircraft tractor. Feature detection, water mist model, image segmentation algorithm and image ranging were used to identify the environmental information in the aircraft tractor motion process and to restrict the motion attitude of the tractor. The transmission protocol based on the 5G data link was designed, the transmission rate and accuracy of control data and video data were verified, and cloud analysis and computing of local data were realized. It was suitable for large scale data unloading, multi-channel HD video transmission and multiple control instruction verification in the outfield test flight.

The book consists of eight chapters. In the first chapter, the aircraft traction cascade system, transmission and control modes of aircraft tractors, aircraft automatic parking and anti-collision system and airport FOD intelligent detection are summarized. The second chapter introduces the development of machine learning, and summarizes machine learning models, data processing and related algorithms involved in the aircraft traction cascade system. The third chapter uses the simulation method to design the function index and parameter index of the aircraft traction cascade system. In the fourth chapter, the PID control and stability of the aircraft traction cascade system are studied. In chapter five, the motion control and environmental constraint control of the aircraft traction cascade system are studied based on machine learning. The sixth chapter introduces the application of high-speed data transmission links in the automatic control of aircraft tractors. In chapter seven, the automatic control of aircraft traction cascade system is simulated. In chapter eight, the research results are summarized, and future development of intelligent automatic control of aircraft tractors is forecasted.

Professor Zhou Yue from Shanghai Jiao Tong University and senior engineer Feng Can from Civil Aircraft Test Flight Center of Commercial Aircraft Corporation of China (COMAC) have given careful guidance in the process of the research. I would like to express my deep gratitude to them. Thanks go to Wu Yijun and other teachers and students from School of Electronic Information and Electrical Engineering of Shanghai Jiao Tong University for their selfless help in the research. I have referred to a large number of literature materials, and I would like to express my thanks to all the authors of the literature.

It is hoped that the publication of this book would promote the development of automatic control technology for large airliner tractors. The author takes all responsibilities for mistakes and errors in this book and suggestions from all readers are highly appreciated.

Liu Tianchang
COMAC, Shanghai
Dce. 2021

CONTENTS

Chapter One

Overview of
Aircraft Traction Cascade System

1. 1 Introduction

Aerospace technology is one of the major features of the comprehensive national strength, and the capacity of research and manufacturing of large civil airliner is an important symbol of national aviation level. At present, China's civil aircraft is in the research and development stage, and there is still a certain distance in design method, manufacturing process and other technologies compared with the international advanced aircraft. The core technology and innovation ability in China's aviation field need to be improved urgently. Aiming at the development of the world's civil airliners, China's domestic airliners have made breakthroughs on the basis of learning advanced manufacturing technologies and adopting equipment of international large airliners. The final assembly of China's large passenger aircraft was finished in Shanghai on November 2, 2015, and its maiden flight was finished successfully on May 5, 2017, which is an important milestone in the history of China's civil aviation industry.

An aircraft tractor is a special airport ground vehicle which is used to tow the aircraft off the airport bridge, onto the runway, into the maintenance or hangar area. The use of tractors can protect aircraft, save fuel and reduce environmental pollution during the sliding of the large aircraft on the ground. The intelligence and flexibility of tractors are particularly important for the aircraft test flights and frequent transfers at different airports.

The process of pushing the aircraft into hangar or towing by a tractor can be abstracted into a machine learning problem by building a machine learning network with image recognition, environment feature detection, motion control information, etc. Machine learning is an interdisciplinary subject involving engineering application, optimization theory, probability theory, statistics, algorithms, etc [1]. Its main research focuses on how to improve existing knowledge structure and performance by learning existing models and methods through multiple methods such as iteration and induction. In this research, machine learning methods were used to complete the simulation of towing and pushing the aircraft by an automatic control tractor. In the process of realizing the automatic control of aircraft traction cascade system by using machine learning, the motion process of the cascade system could be abstracted into a mathematical problem. The simulation control of the cascade system was realized by the model data processing, parameter optimization and constructing the supervised learning model with the machine learning network.

The internal links of the cascade system are extremely complex. Many parameters such as bias angle, traction direction, and tire steering have a non-linear relationship with the motion state of the system while pushing the aircraft into the hangar. The state of the system can not be corrected after the cascade system exceeded the threshold angle. The different motion states of the cascade system will pose new challenges to the existing control methods. In the research, a machine learning model based on a large number of data samples was proposed to achieve accurate control of the cascade system under nonlinear and multi-state conditions.

In the test flight of civil aircraft, a combination of multiple sensors is often used to enable aircraft tractors to obtain the surrounding environmental information. Due to the large number of video signal inputs, high sampling frequency, and the limitation of the tractor's volume, it is difficult to directly perform machine learning on the tractor and complete the data analysis of the control process. In the actual traction process, the economic benefits affected by the tractor's speed need to be considered, and the moving speed of the tractor should be as high as possible while maintaining stability. The acquired environmental information can be directly sent to the data-center by wireless transmission technology, the data analysis based on machine learning can be completed by data-center's super computing power, and the control commands can be issued [2]. For the analysis of aircraft tractors, 1080p (Progressive Scanning) cameras with more than 8 channels and a single channel of more than 5M are needed to obtain environmental information, and multiple sensors are used for data transmission and processing. The total bandwidth requirement is above 600M, and the delay is extremely low, which is difficult to be met by existing technologies. High-speed communication link is composed of automatic control unit, high-speed full-band transmitter, high-speed full-band receiver and data center. The current data of the aircraft tractor is obtained by the automatic control unit, the data is sent from the high-speed full-band transmitter to the receiver. The receiver transmits the data to the data-center after data transcoding, the data is analyzed quickly in the data-center, and the control instruction is sent to the aircraft tractor, which executes the control command. For the large-scale data, specific transmission methods, data verification technology, antenna self-alignment technology, and data encryption technology need to be studied. The interface design of high-speed data transmission software has become an urgent problem to be solved in order to ensure the stable transmission of large bandwidth data in high-speed communication network [3-4] .

At present, there are some problems in the test flight of large passenger aircraft, such as the difficulty of tractor supporting in the outfield base, long period of personnel training and low rate of data transmission. There are few researches on the automatic control technology of aircraft tractors all over the world, and high-speed data transmission technology is rarely used in the aviation industry. The airport image recognition technology is still in the aircraft type recognition, and the research on the overall environmental information image recognition is rare, the database is mainly aircraft model data, lacking data of various parts of the airport.

In summary, this study aims at the test flight needs of China's passenger aircraft and carries out the intelligent simulation design of aircraft tractor through machine learning algorithm and high-speed technology, and the automatic control of the multi-cascade and nonlinear system was studied in the process of pushing the aircraft into storage. The automatic and autonomous control of the aircraft tractor was realized by simulation. The application of machine learning, high-speed data link and intelligent image processing in the automatic control of aircraft tractors expand the application fields of the machine learning and intelligent image processing, which is of great significance for improving the efficiency of test flight and the safety of tractors.

1.2 Transmission and control modes of aircraft tractors

The transmission control systems of aircraft towing tractors include mechanical, electric, hydrodynamic and hydraulic transmission [5-7] . The mechanical transmission was used in early aircraft towing tractors, but it has been rarely used because of its poor flexibility and other shortcomings. Electric power transmission is mainly used in small and medium-sized aircraft towing tractors due to the influence of transmission power and the battery life. Hydraulic transmission is mainly used in the aircraft towing tractors; it is connected by flexible pipe, and it has a flexible structure, a wide range of stepless speed regulation, good dynamic braking characteristics superior to the ability of overload protection, and a continuous transformation output [8-9] .

Lufthansa and Krauss Maffei manufactured the world's first towbarless aircraft tractor in the early 1980s. So far, the more famous are Douglas company in UK, GHH company in Germany, TLD company in France, FMC company in the United

States, IAI company in Israel and so on. Some manufacturers in China have begun to produce aircraft tractors since 1990s. At present, some Chinese enterprises in Beijing, Qingdao, Shenzhen, Shenyang, Weihai and other cities can produce small and medium-sized aircraft towing tractors.

1.2.1 Traction modes

According to the connection between the aircraft and the aircraft towing tractor, the traction mode of aircraft towing tractor can be divided into towbar traction and towbarless traction. Towbar traction has been applied in civil aviation airports for a long time. The aircraft and the tractor are connected by the traction bar to control the movement of the aircraft under the push or pull action of the traction bar. The traction bar has torsion shear and buffer device, which can transfer traction force and reduce the impact between the tractor and the aircraft. The towbarless tractor cancels the traction rod and uses its clamping lifting device to directly connect with the landing gear of the aircraft. At present, it has developed rapidly and has become the main traction of the aircraft towing tractor [10-12]. The relationship between the aircraft towing tractor and the aircraft is shown in Figure 1 - 1.

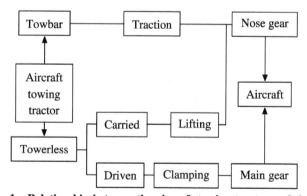

Figure 1 - 1   Relationship between the aircraft towing tractor and the aircraft

By comparing the interaction between the aircraft towing tractor and the aircraft, the characteristics of towbar and towbarless traction are shown in Table 1 - 1.

Table 1 – 1   Characteristics of towbar and towbarless traction

Type of traction	Advantages	Disadvantages
Towbar	(1) Structure, control and drive are simple and flexible, low production cost. (2) The tractor is not directly contacted with the aircraft, so it is not necessary to consider the adaptability of the connection device. The aircraft can be pulled as long as there is a corresponding traction bar and appropriate traction force. (3) It can be used to pull aircraft and also be used to carry other equipment through the connection of the traction bar.	(1) In the traction process, the turning radius is large and the traction speed is low. The installation and disassembly of the traction towbar need a specially-assigned person, the operation is complex and the working efficiency is low. (2) Different types of aircraft should generally be equipped with different traction towbars. The application scope is limited. (3) The tractor is heavy and the energy consumption is high.
Towbarless	(1) The overall length of the unit is shortened, the traction turning radius is small, and the traction speed is fast. (2) The tractor is light, so energy consumption is lower. (3) The traction process can be completed by the driver of the tractor, because the operation is easy.	(1) The structure is complex, special transmission and control components are required, the cost is high. (2) Only specified aircraft can be towed, the scope of application is limited. (3) The towbarless tractor shall be equipped with a set of traction limiter to prevent the aircraft from being damaged due to excessive traction and braking force on the landing gear.

From the development trend, towbarless traction will become the main form of aircraft towing tractor. The main characteristic of the towbarless aircraft tractor is that hydraulic drive is used and the main power source is the internal combustion engine, amd it can meet the needs of multi-type aircraft traction. In order to meet the comprehensive guarantee of various working conditions, aircraft tractor will develop towards the direction of high efficiency, energy saving, emission reduction and intelligence. In the future, aircraft tractor can be designed into highly integrated intelligent products such as machine, electricity, liquid and gas through intelligent vehicle technology[13]. In addition, the aircraft is automatically towed to the corridor bridge by an unmanned aircraft tractor after landing through unmanned driving technology[14] , and it can be directly sent to the runway by an unmanned vehicle when the aircraft leaves the tarmac, which can save a lot of fuel, improve the utilization rate of the vehicle, and reduce energy consumption. With the continuous expansion of airport construction, it has a broad prospect to develop a new generation of aircraft tractor combined with the intelligent automobile and unmanned driving technology.

1.2.2   Transmission control system

There are four main transmission control systems for aircraft towing tractor, i. e. mechanical, electric, hydrodynamic and hydraulic. Mechanical transmission has been rarely used due to the impact of shifting, large parts, cumbersome size and other reasons. Because some key parts of electric transmission have not reached the practical application level, it is mainly used in small tractors, and will be applied to medium tractors in the future. Hydraulic transmission is gradually becoming the mainstream transmission mode of the large and medium-sized aircraft towing tractors due to the wide and infinitely variable speed range, better power braking characteristics, superior overload protection and the ability to continuously transform the output power[15-16] . The characteristics of the four transmission control systems are shown in Table 1 - 2.

Table 1 - 2   Characteristics of aircraft towing tractor transmission control system

Transmission control system	Advantages	Disadvantages
Mechanical control system	(1) Simple structure, mature technology, fewer precision parts, lower price. Maintenance and repair is easy. (2) High transmission efficiency, low failure rate. The change of output speed with the load is small, and the requirements for operation conditions are not strict. (3) Strict match with the engine, which can be used to brake in a certain speed range.	(1) Gear transmission has shifting impact, and starting is not gentle, which is adverse to smooth traction. Shifting is accompanied by power interruption, so the efficiency is low for vehicles with frequent shifting. (2) A longer pull rod is usually used, overall layout is limited, and the scheduling is not flexible, operation is cumbersome. It is difficult to realize remote control and automatic control. (3) The load varies greatly. It is easy to cause a sudden change in engine speed. The engine is easy to stall when overload is running.

(Continued)

Transmission control system	Advantages	Disadvantages
Electric control system	(1) The motor and generator are connected by flexible cable, which is more convenient than hydraulic pipe. The association degree with microelectronic technology is high, and it is easy to achieve automatic control and remote control. (2) Stable engine operation, low fuel consumption, small pollution and long life. (3) Easy to realize internal combustion engine and battery hybrid mode drive.	(1) Power density is small, so it is mainly used for small towbarless tractors. (2) The motor is large and heavy, and the structure is very cumbersome and complicated. (3) Nonferrous metals and other special materials are consumed more than others, so the price is higher. (4) It is difficult to solve the problems of anti-jamming and electromagnetic compatibility.
Hydrodynamic control system	(1) The output speed and torque can be maintained within a certain range. The output torque can be automatically increased when the load increases. (2) Automatic gearbox is equipped, operation conditions are significantly better than mechanical transmission. (3) The shifting number of times is obviously smaller than that of the mechanical transmission, and the invalid time of power interruption is shortened. (4) The torque converter can absorb the impact load, avoid the overload, and extend the working life of the engine.	(1) The torque of hydraulic converter is small, and the sudden change of gearbox transmission ratio still exists. The matching between torque converter and engine is very demanding, so the versatility is poor. The structure is more complicated than the mechanical transmission and the cost is higher. (2) The maximum output torque can only be obtained when the engine speed is the highest, but the efficiency is very low when the engine speed is low and the torque is high, which is not conducive to energy saving, emission and noise reduction. (3) The power braking ability is poor, and the driving brake works frequently and wears quickly. (4) It is difficult to arrange the structure in the towbarless tractors with four-wheel drive.

(Continued)

Transmission control system	Advantages	Disadvantages
Hydraulic control system	(1)The input and output components are connected by flexible pipes. Starting, speed regulation and reversing are soft and fast, and the micro feature is excellent. (2) The traction force and the speed adjustment characteristics are good, and the vehicle can produce high output torque at low speed. The engine can run in stable and efficient conditions, which is beneficial to energy saving, emission reduction and engine life extension. (3) Controllable braking can be achieved by power transmission from maximum speed to full rest. The braking torque is equivalent to the driving torque, which is less likely to cause wear of the components.	(1) Precision components, high manufacturing technology, high cost. (2) Steady-state transmission efficiency is low, which is not suitable for long distance stable driving vehicles. (3) The maintenance is more complex, and the maintenance technical level is high.

1.2.3   Hydraulic transmission control system of aircraft towing tractors

i . Hydraulic steering control system

The steering control system of aircraft towing tractor includes front wheel steering, four-wheel centripetal steering and four-wheel crab steering. The steering system is mainly composed of steering pump, steering device, electromagnetic reversing valve, constant flow valve, hydraulic control one-way valve and steering cylinder. In order to keep the steering oil pipe full of oil, hydraulic control one-way valve is used to guarantee the synchronous movement of steering wheel. The front and rear wheels can be returned to the center position through the resetting device to ensure the correct driving direction of the vehicle.

According to the front and rear axle steering cylinder connection methods, hydraulic steering system can be divided into the series system and the parallel system. The series system mainly consists of hydraulic pump, steering device, front axle steering cylinder, rear axle control valve group and rear axle steering cylinder. The rear wheel steering responds faster because the oil in the front axle steering cylinder does not return to the hydraulic tank but directly reaches the rear axle steering cylinder. However, the series steering may have the problem of lag of the rear wheel steering due to the leakage of hydraulic components at different degrees. The parallel system is mainly composed of hydraulic pump, steering device, steering mode control valve group, two-way overload valve group, front axle steering cylinder and rear axle steering cylinder. When the steering control system works, the hydraulic oil flows through the steering mode control valve group, and then flows to the front axle steering oil cylinder and the rear axle steering oil cylinder independently through the shunt valve. The return oil circuit of the front axle steering oil cylinder is directly connected to the oil tank and is relatively independent.

ii. Hydraulic braking control system

The hydraulic braking control system of aircraft towing tractor is mainly composed of hydraulic pump, one-way valve, electromagnetic reversing valve, pressure switch, accumulator, manual pump, hand brake valve, foot brake valve, shuttle valve and relay valve. The brake safety can be guaranteed by double circuit brake valve. The two spool valves in the double circuit brake valve are mechanically connected and protected by two accumulators. When braking is completed, the oil returns to the tank through the brake and relay valve. If one of the braking circuits fails, the other can continue to operate to ensure braking safety.

With the development of automobile intelligence, the traditional brake system can not meet the requirements of the vehicle control system, so the line control technology has developed rapidly, such as automatic emergency braking system and advanced driving assistance system. Line control system can be divided into electromechanical braking system and electronic hydraulic braking system. Electronic hydraulic braking system is an important development direction of the automobile braking system, and its main characteristic is to replace some mechanical parts in traditional braking system with electronic components. Electronic hydraulic braking system retains the original mature and reliable hydraulic parts, with the advantages of compact structure, rapid response, accurately controlled braking force, and it is easy to achieve a variety of active safety control functions. Many well-known vehicle and parts companies, such as German Bosch, Continental LSP, Daimler Chrysler, American TRW, Delphi, Japanese Hitachi, Honda, Toyota, Korean Mando, etc. , have put forward the electronic hydraulic braking system scheme and its hydraulic pressure control method, and carried out simulation and experimental research. Tsinghua University, Jilin University, Tongji University, Nanjing University of Aeronautics and Astronautics, Wuhan University of Technology, Jiangsu University and others in China have carried out simulation or bench test research on the electronic hydraulic braking control system.

The main factors affecting the energy consumption characteristics of electronic hydraulic braking control system include braking intensity, the maximum and minimum working pressure, air pressure of accumulator, accumulator effective displacement and brake wheel cylinder piston diameter. The energy consumption of electronic hydraulic braking system is affected by the strength of braking and braking wheel cylinder piston diameter. Reducing the maximum working pressure of the system and the piston diameter of the braking wheel cylinder is beneficial to reducing the power consumption of the electronic hydraulic braking system. The change of the minimum working pressure of the system and the effective displacement of the accumulator has little influence on the energy consumption of the electronic hydraulic braking system. Increasing the charging pressure of the accumulator, reducing the effective displacement of the accumulator and reducing the piston diameter of the brake wheel cylinder are beneficial to reducing the volume of the accumulator.

iii. Hydraulic clamping and lifting mechanism

Clamping and lifting mechanism can be divided into clamping mechanism and lifting mechanism. It is the part directly connected with the towbarless aircraft tractor, and it is also the key mechanism of the towbarless aircraft tractors. Clamping mechanism includes four forms, i. e. rotary clamping type, symmetric orbital type, asymmetric orbital type and orbital spring type. Symmetric orbital type and rotary clamping type are more commonly used. German Goldhofer company produces special ground equipment for airports, and its towbarless tractor mainly includes AST-1 and AST-3 series. AST-1 adopts symmetrical rotary clamping mechanism and AST-3 adopts symmetrical track clamping mechanism.

The clamping and lifting mechanism of towbarless aircraft tractor is mainly composed of clamping mechanism, bearing lifting platform, hinge head, hydraulic cylinder and other components. When the aircraft is docked with towbarless tractor, the aircraft landing gear enters the clamping lifting device of the tractor by the hydraulic cylinder lifting action. The front wheel of the aircraft is completely fixed with the lifting platform by the hydraulic clamping device, and is clamped and lifted to a certain height by the lifting device which is driven by the lift hydraulic cylinder.

In order to adapt the using of A380, Douglas began to develop the TBL-600 aircraft towing tractor in 2005. Its clamping mechanism adopts the single-side rotary clamping. The German SPACER series hydraulic system provides constant pressure support to the left and right wheels of the front landing gear, this configuration serves as a safety protection for the nose landing gear. With the improvement of the automation of the towbarless aircraft tractor, Schopf has developed a semi-automatic aircraft tractor, adopting the driving clamping mechanism. During the process of clamping and traction, the aircraft wheel always keeps contact with the ground and rolls. Guangtai company has proposed a new type of towbarless tractor, which can restrain the front landing gear wheel at least three points at the same time when clamping the aircraft front landing gear, thus improving the safety of traction.

iv. Hydraulic hybrid control system

Hybrid electric vehicles can recover energy and store it during braking, and the energy is released during starting and accelerating, thus achieve the purpose of energy recovery and reuse. This is the energy saving mechanism of hydraulic hybrid system: when the tractor needs to brake or slow down, the secondary element (hydraulic pump/motor) works in the state of hydraulic pump, with the braking energy of the tractor stored in the hydraulic accumulator in the form of high pressure hydraulic oil by compressing the gas in the accumulator, so the vehicle's kinetic energy is converted to the form of hydraulic energy stored in the accumulator. When the tractor starts up again and accelerates, the accumulator releases energy, and the hydraulic oil flows through the secondary components to drive the tractor, so that the energy is converted from the form of hydraulic energy to the kinetic energy, thus reducing the engine fuel consumption.

Compared with the electric hybrid vehicle, the hydraulic hybrid vehicle overcomes the problems of low energy recovery and battery affected by temperature, so it has the advantages of long life, high reliability, easy maintenance, simple structure, high power density, flexible layout and so on. More and more attention is being paid to the hydraulic hybrid vehicle with the development of vehicles towards energy saving and emission reduction.

Hydraulic hybrid power is a new type of power system based on secondary regulation hydrostatic transmission technology. The hydraulic hybrid power system can be divided into the parallel type and the serial type according to the connection between the main power source and the secondary components. The comparison between the parallel type and the serial type is shown in Table 1 − 3. It can be seen that the overall performance of the serial type hydraulic hybrid power system is superior to that of the parallel type. By combining and changing the two basic structures of the serial and the parallel, the mixed structure and wheel side driven structure can be obtained.

Table 1 − 3   Comparison of parallel and serial hydraulic hybrid power system

Type	Engine efficiency	Kinetic energy recovery efficiency	Energy conversion efficiency	Control performance	Weight
Parallel	The operation mode is almost the same as traditional vehicles, and the overall efficiency is limited.	Parallel and serial are the same energy recovery system, and the recycling efficiency is basically the same.	Mechanical transmission is still needed, and energy conversion efficiency is low.	The engine system and the hydraulic system need to be controlled separately, which is more complicated.	Heavy
Serial	The vehicle engine is only used as the power output to drive the hydraulic pump, which has higher efficiency.	The hydraulic system provides power output for the vehicle, and the power is high.	The energy conversion is completed by hydraulic system, and the conversion efficiency is high.	The hydraulic system is connected with the vehicle engine; the control is relatively simple.	Light

1.2.4   Automatic control of aircraft tractors

Some relevant researches have been carried out on the automatic control algorithm[17−18], but the aircraft traction still needs manual control on the ground at present. For the airport with high security requirements, such as some large international airports, the pure tracking real-time control algorithm has unique advantages, and it can deal with various emergency situations. The core concept of the aircraft tractor automatic control technology is to change the current way that the tractor driver controls the aircraft traction, which can realize that the intelligent towing system can tow aircraft on the ground under all conditions autonomously, or the aircraft can be remotely controlled by the pilot, thereby transforming the tractor from an independent device into an intelligent auxiliary power device for aircraft taxiing on the ground. The new intelligent aircraft tractor is the development of the traditional towbarless tractor. The intelligent system design improves the operation flexibility and accuracy of the tractor, and guarantees the aircraft scheduling and safety.

When the automatic control technology is applied to the control of aircraft tractor, the operator can control the tractor wirelessly by portable digital control system[19]. The operator can observe the tractor's movement intuitively, so as to improve the accuracy of operation and avoid accidents during the remote control. There are a variety of sensors installed on the smart aircraft tractor, which can be combined with GPS and other positioning data to calculate the moving position of the tractor. The data communication module can transmit the real-time images and the data collected by sensors to the computing center in the form of data streams, and the data is parsed and sent back to the tractor to realize automatic control. Also, this kind of tractor can identify and avoid obstacles encountered during the motion, so as to realize automatic protection.

1.3 Aircraft automatic parking and anti-collision system

The frequent takeoff and landing of aircraft will increase the number of vehicles in the airport, especially in the process of towing aircraft in or out of the hangar. A large number of vehicles are likely to cause collisions in the airport between aircraft and vehicles or buildings, and it affects the safety of passengers and aircraft seriously. The tractor driver cannot visually detect the distance between the fuselage, wings and adjacent obstacles due to the irregular shape of the aircraft and the viewing angle. Many ranging methods and real-time signal processing can be applied to the aircraft parking anti-collision system with the development of digital signal processing technology[20]. Nowadays, the anti-collision warning technology has a wide range of applications, and can be used in various transportation fields, such as civil hangars, military hangars, automobiles and ships[21-24].

Aircraft anti-collision system is a tractor driving safety assistance technology based on the active safety protection. According to obstacles' information, the blind spots of the driver can be eliminated when parking and starting the tractor by distance measurement, sound or more intuitive display, thus the driving safety can be improved. The United States, Japan, the European Union and other countries have increased their researches and development efforts in many kinds of ranging equipment in the aviation field. At present, various advanced electronic ranging technologies have been developed rapidly, and ranging technologies have continued to innovate.

The aircraft automatic parking system plays an important role in the airport information construction[25-26]. As the plane is taxiing from landing to parking, pilots can obtain runway and hangar information, such as taxiing speed of the aircraft, distance to the hangar, whether deviating from the taxiing centerline, through the aircraft's automatic parking guidance system. At the same time, the automatic parking system can also collect the aircraft's docking state, docking time, docking location and real-time environmental information, and upload relevant information to the central database for centralized processing and rapid statistical analysis[27-28].

The aircraft automatic parking system of using laser inspection or visual inspection technology has different characteristics[27-28]. The aircraft position detection of large-angle non-contact technology can provide information for the aircraft automatic parking navigation, the aircraft automatic parking system of using laser scanning technology can perform relevant detection of aircraft width, height and length, which can provide real-time navigation information for aircraft parking while using visual inspection and laser inspection, such as the use of APIS technology for laser detection.

There are few researches on the aircraft automatic parking system in China. University of Electronic Science and Technology of China and Civil Aviation University of China have carried out some researches in this field[29-31]. Some international airports have installed the aircraft automatic parking system, but the guidance is not accurate enough by the automatic parking system, and manual navigation is still required.

1. 4   Airport FOD intelligent detection

According to the definition of the Safety Technology Center of the State Civil Aviation Administration of China in 2009, Foreign Object Debris (FOD) refers to any foreign object that does not belong to the airport but appears in the operation area of the airport and may cause loss to the airport or damage to the aircraft. FOD includes discarded metal parts, plastic fragments, rocks, concrete fragments, rubber fragments, tree branches and all kinds of animals breaking into the airport. FOD can cause serious damage to aircraft engine blades, airframes and wheels due to the strong suction power of the aircraft engine and the high-speed friction between the tire and the ground during takeoff and landing. It will also cause casualties, flight delays, interrupted takeoffs, runway closures and other issues. FOD inspection is an important part of the safe operation of aircraft and has important significance for flight safety. With the development of the flight test of China's large passenger aircraft, the aircraft needs to carry out flight test and data collection in a number of airports in different environments. For example, the flight test is conducted under the environment of high temperature, high humidity, natural icing, strong wind, plateau and high cold. High-intensity flight tests take place frequently at different airports, and it puts forward new requirements and challenges for the FOD intelligent identification.

Traditional FOD detection relies on the manual intermittent inspection, which has high labor intensity and low detection efficiency. The method of manual inspection cannot meet the needs of flight safety. With the development of the real-time video technology and computer image recognition technology[32-34], the technology of FOD automatic and intelligent recognition has been developed rapidly. For example, the method of millimeter-wave ground synthetic aperture radar imaging is used for FOD detection[35-37]. The ISAR/SAR imaging technology is used to eliminate the clutter in the complex background of the airport with the CFAR detector[38], and FOD is classified in the feature domain with SVDD classifier[39]. The X-band vehicle-mounted synthetic aperture radar can be used to FOD detection at lower frequencies[40]. Some researchers use the method of deep convolutional neural network to FOD recognition[41-42], and FOD can be identified by the methods of high-resolution radar[43], particle tracking[44], and machine learning[45]. The FOD intelligent detection system can deal with the shortcomings of manual inspections. Some problems, such as insufficient runway inspection time caused by dense flights in large airports during rush hours, have been alleviated. The intelligent detection system can be used to monitor and detect FOD all-weather, and it gives alarm in time and improves the efficiency of FOD processing.

Some airports use the intelligent FOD monitoring system as a auxiliary method of FOD detection at present, and the real-time detection of FOD can be performed to achieve precise positioning and timely alarm. The intelligent detection method can effectively reduce the safety hazards caused by the FOD. Some common FOD monitoring systems are American FOD Finder system, British Tarsier Radar system, Israeli FOD detect system and Singapore iFerret system. For the four systems, the iFerret system uses a combination of real-time image video and millimeter wave radar for FOD detection, and the other three systems use millimeter wave radar for FOD detection.

China's awareness of FOD recognition has gradually increased, and the Civil Aviation Administration of China issued regulations on the airport FOD testing in 2009. The first domestic FOD monitoring system was installed and used in Xiaoshan Airport in 2016, so the airport staff can control the system and carry out some relevant operations through the system program.

There are many FOD intelligent detection algorithms[46-49]. The image segmentation algorithm, such as binary detection, grayscale detection, active contour detection and OTSU detection, can detect and recognize the image contour[50]. For the outfield flight test, the FOD detection and alarm system cannot be constructed in time, but the FOD contour detection based on the image recognition can be used to detect and recognize the FOD[51]. The binarization method firstly transforms the acquired color image into a gray image, and then adjusts the gray value of pixels on the image to 0 or 255 according to the threshold value. The whole image appears in black and white colors, which greatly reduces the amount of data in the image and makes the main component of the target contour visible[52-53]. The OTSU method calculates the variance of the image foreground and background according to the gray characteristics[54], so as to obtain a better image contour recognition area.

Chapter Two

Machine Learning
Models and Algorithms

2. 1　The development of machine learning

Machine learning has been developed for more than half a century since its inception. The development has the following characteristics:

(1) The artificial neural network based on the neuron model has experienced many ups and downs [55] , running through each stage of the development of machine learning.

(2) The proposal of new algorithms in machine learning is closely connected with the learning theory [56] . Nevertheless, the current theory of machine learning is not perfect, and various machine learning models, including artificial neural networks, are still classified as "black box decision-making" .

(3) In specific applications, machine learning is mainly used for clustering and classification, and there are relatively few applications in prediction. This is related to its poor interpretability and low stability. Of course, this conclusion is mainly based on the limited data. In the era of big data, machine learning will make new breakthroughs when dealing with prediction problems.

(4) The application of machine learning in the financial field is not deep enough compared to the wide application in industry information. Machine learning will play a greater role in big data risk control and other fields with the rapid development of new finance.

2.1.1　Budding period

The first stage was the budding period in the 1940s. The main achievements of this stage are Bayesian theory, least squares method, Markov model and neuron model.

i . Bayesian theory

Bayes' theorem is a theorem about the conditional probability of random events A and B in probability theory. It describes the probability of an event occurring under certain conditions. For example, if it is known that the performance of a certain airframe structure is related to the number of takeoffs and landings, Bayes' theorem can be used to calculate the performance of the airframe structure through the known age of the aircraft, that is, the approximate number of takeoffs and landings, so as to determine whether the structural parts need to be maintained or replaced.

Generally, the probability of event A occurring under the condition that event B has occurred is different from the probability of event B occurring under the condition that event A has occurred. However, there is a definite relationship between the two events, and Bayes' theorem is a statement of this relationship. One use of the Bayesian formula is to deduce the fourth probability through the three known probabilities. Bayes' theorem is related to the conditional probability of the machine variable and the marginal probability distribution. It can also be used for continuous probability distributions.

ii. Least squares method

Least squares method is a mathematical optimization modeling method. It is developed in the fields of astronomy and geodesy. It was proposed in the middle of the eighteenth century. Scientists and mathematicians tried to provide solutions to the ocean navigation challenges in the period of great navigation exploration. It seeks the best data by minimizing the square sum of errors.

In the least squares method, the combination of different observations is the best estimate of the true value. Multiple observations will reduce the error rather than increase it. It is a standard method to obtain approximate solutions through regression analysis. In the entire solution, the least squares method is calculated to minimize the sum of the squares of the residuals in the result of each equation.

When evaluating the calculation method for minimizing the error, Laplace pointed out the mathematical form of the probability density of the error and defined the estimation method for minimizing the error. For this reason, Laplace used a bilaterally symmetric exponential distribution, that is, an error distribution model, and used the sum of absolute deviations as the estimation error. The least squares method used by Gauss was published in *Theory of Celestial Movement* in 1809. In 1829, Gauss provided proof that the optimization effect of the least squares method is stronger than other methods, namely the Gauss-Markov theorem. The least squares method can be used to easily obtain unknown data, and minimize the sum of squared errors between the obtained data and the actual data.

iii. Feedforward neural network

Feedforward neural network is the simplest kind of neural network. It adopts a unidirectional multilayer structure. Each layer contains several neurons. Each neuron is arranged hierarchically. Each neuron can receive the signal of the neuron of the previous layer, and generate output to the next layer. There is no feedback between layers. The zeroth layer is called the input layer, the last layer is called the output layer, and the other intermediate layers are called hidden layers. The hidden layer can be one or multiple layers.

The research on the feedforward neural networks began in the 1960s, and the current theoretical research and practical applications have reached a very high level. A typical multilayer feedforward neural network is shown in Figure 2 – 1.

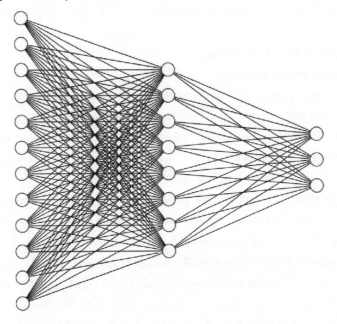

Fig. 2 – 1   Multilayer feedforward neural network

iv. Hidden Markov model

Hidden Markov model is a type of statistical model used to describe a Markov process with hidden unknown parameters. The difficulty of analysis is how to determine the hidden parameters of the process from the observable parameters and make further analysis with these parameters.

In the normal Markov model, the state is directly visible to the observer, so the transition probability of the state is all the parameters. In the hidden Markov model, the state is not directly visible, but some variables affected by the state are visible. Each state has a corresponding probability distribution on the possible output symbols, so the output sequence of the object can reveal some information about the state sequence.

Hidden Markov model is commonly used in the field of pattern recognition, such as speech recognition, Chinese word segmentation, optical character recognition, machine translation, bioinformatics and genomics, prediction of protein coding regions in genome sequences, the process of decoding in communication, and the map matching algorithm.

The pattern recognition is a mathematical method used to study patterns automatic processing and interpretation by computer [57-58]. If the sample is complex and discrete, this algorithm should not be used [59-62]. The K-nearest method is not used to predict the samples of the aviation system because the samples are not strong aggregation and they are usually scattered.

v . M-P neuron model

M-P neuron model originated from a groundbreaking paper published in 1943. The two authors of the paper are neurophysiologist Warren McCulloch and mathematician Walter Pitts. The paper describes the neuron time summation, domain value and other characteristics of neurons. For the first time, it realizes using a simple circuit to simulate the behavior of brain neurons.

In this model, each neuron can receive signals transmitted from other neurons. These signals are often weighted, and then compared with the threshold value inside the receiving neuron, and the corresponding output is generated through the neuron activation function.

The M-P neuron model has the following characteristics:

(1) Each neuron is an information processing unit with multiple input and single output.

(2) There are two types of connections between neurons: which are excitatory and inhibitory. When a neuron is in an excitatory state, it will send a signal to the connected neuron and change its potential.

(3) Each neuron needs to integrate all input signals and decide whether to be excited according to the threshold, that is, the neuron has spatial integration characteristics and threshold characteristics. When the potential of the neuron receiving the signal exceeds its own threshold, it excites and repeats the signal sending process.

(4) The selection of activation function should depend on the specific application, which is mainly divided into continuous type and discontinuous type.

The M-P neuron model is the basis of neural network learning, and the model is the earliest and longest used model in machine learning[63].

2.1.2 Enthusiastic period

The second stage was the enthusiastic period from the mid-1950s to the mid-1960s. The operation process of neurons was clarified in the embryonic stage, but the efficient operation of neural network learning needs to rely on the relevant learning rules. The sign of the enthusiastic period is the introduction of the classic learning rules, such as Hebbian rule, perceptron model, two classification model, gradient descent model, least squares model.

i . Hebbian rule

In 1949, psychologist Hebb put forward the hypothesis of synaptic modification related to the neural network learning mechanism. The core idea is that when two neurons are in an excited state at the same time, the degree of connectivity will increase. The weight adjustment method based on this assumption is called the Hebbian rule. Since the Hebbian rule belongs to the unsupervised learning, it has limitations when dealing with a large number of labeled classification problems.

Hebbian theory explains how neurons form connections to make memory marks. Hebbian theory has also become the biological basis of unsupervised learning. Hebb clarified the morphology and function of neurons: "If two neurons or the neuron system is always excited at the same time, it will form a kind of combination in which the excitation of one neuron will promote the excitation of the other." Hebb also wrote: "If one neuron continues to activate another neuron, the axon of the former will grow out of the synaptosome and connect with the cell bodies of the latter; if they already exist, they will continue to grow."

Gordon proposed the role of cell clusters and their role in the formation of memory traces based on the idea of auto-association: "If the input of the system will cause the same pattern to reappear, then the interrelationship between the elements that make up this pattern will be greatly enhanced. This means that any one of the elements will tend to trigger other elements in the same group, and at the same time, it will inhibit the other unrelated elements in a way of reducing weight. From another point of view, this model realizes self-association, and the model that has been learned can be called a memory trace, that is, a self-association model."

It is generally believed that, from a holistic perspective, Hebbian theory is the primary basis for neural networks to form memory traces.

ii . Perceptron model

In 1957, in the Cornell Aeronautical Laboratory, the American neuroscientist

Rosenblatt proposed the simplest forward artificial neural network, which is perceptron model, and successfully completed the simulation of the perceptron on the IBM 704 machine. Two years later, he successfully realized Mark 1, a neural computer based on a perceptron that can recognize some English letters. It was shown to the public on June 23 in 1960, which started the first supervised learning. In 1962, Novikoff deduced and proved that, when the sample is linearly separable, the perceptron can be converged after a finite number of iterations, which provides a theoretical basis for the application of perceptron learning rules.

The perceptron is composed of two layers of neurons, and its output is a single M-P neuron. The activation process in the entire network is only completed in the output layer. The input layer receives data and transmits it to the output layer. It does not have the activation function, resulting in very limited functions of the model. When dealing with simple linearly separable problems, the calculation can be guaranteed to converge and solve with the appropriate weight.

The biggest feature of the perceptron is that it can solve the binary linear classification problem through iterative trial and error. When the perceptron was proposed, solving algorithms were also born, including perceptron learning method, gradient descent method and least squares method. At present, perceptrons are widely used in the recognition of text, sound, signal, learning memory and other fields.

iii. Binary classification model

The binary classification model is often used to judge the classification problem, for example, whether aircraft tires meet the manufacturing requirements through the results of material analysis. A classifier is trained to input the material analysis results of the aircraft tire, which is represented by the feature vector X, and output the availability of the tire, and the output is represented by the corresponding value. For different input feature vectors, the mapping results are correspondently different. This is a typical binary classification problem.

In practical applications, the recognition rate is usually calculated in the form of probability in order to effectively classify the results. For example, a probability of 90% means that the tire can continue to be used, and a probability of 10% means that the tire can no longer be used, then a threshold is set and the results are classified. This requires the accuracy of the classification results to be between 0% ~ 100%, but the general linear model cannot make the recognition results between 0 and 1. In order to map all the ordinate values to the 0 ~ 1 interval, the logistic model is usually used to map the classification results and effectively classify them.

iv. Least squares model

In 1809, Gauss clarified the method of least squares in *Theory of Celestial Motion*, and in 1829, he provided that the optimization effect of the least squares method is stronger than other methods.

The least squares method is a mathematical optimization technique that finds the best function of the data by minimizing the square sum of the error. It can be used to obtain the unknown data easily and minimize the sum of squares of the errors between the obtained data and the actual data. It can also be used for curve fitting, and some other optimization problems can be expressed by the least squares method.

2.1.3   Calm period

The third stage was the calm period from the mid-1960s to the mid-1970s.

Since the perceptron has a single structure and can only handle simple linear separable problems, how to break through this limitation has become the focus of machine learning theorists. Although it was initially considered to have good development potential, the perceptron eventually proved unable to deal with many pattern recognition problems.

In the calm period, the development of machine learning has almost stagnated. The main reasons are as follows:

(1) Lack of theory is a key factor of restricting the development of artificial neural networks.

(2) With the increasing difficulty of real-world problems, the application limitations of single-layer artificial neural networks are increasing. Although Winston's structural learning system and Roth's logical induction learning system appeared in this period, they were not put into practical use because they could only learn a single concept.

(3) The limited memory and slow processing speed of computers restrict the application of machine learning algorithms. At the same time, the capacity of the database was relatively small, and the increase in the size of the data also distorted the effect of a single machine learning algorithm.

(4) A group of scholars represented by Minsky and Papert have seriously questioned the effect of the perceptron. Through rigorous derivation and publication

of works, such as *The Perceptron* published in 1969, they proved that the perceptron could not solve linear inseparable problems, such as simple exclusive OR(XOR), and even the most advanced computers at that time did not have enough computing power to complete the huge amount of computation required by neural network models. After that, many countries stopped funding neural network research, which further accelerated the decline of single-layer artificial neural networks with the perceptron as the core.

At this point, the research of artificial intelligence had entered a low tide of nearly two decades, which is called AI Winter in history.

2.1.4   Revival period

The fourth stage was the revival period from the mid-1970s to the late 1980s.

In 1980, the first international symposium on machine learning was held in the United States by Carnegie Mellon University, which marks the rejuvenation of machine learning all over the world.

In 1983, Hopfield, a physicist at the California Institute of Technology, used a new type of fully interconnected neural network to solve the traveling salesman problem.

In 1986, the publication of *Machine Learning*, a professional journal in the field of machine learning, meant that machine learning became the focus of theoretical and industry attention once again. During the revival period, the biggest breakthrough in the field of machine learning was the variety of artificial neural networks, which made up for the shortcomings of the single structure of the perceptron.

In 1986, Rumelhart and McClelland of UCSD co-authored the book *Parallel Distributed Processing: Exploration of Cognitive Microstructure*, in which a learning rule applied to multi-layer neural networks (BP algorithm) was proposed, which promoted the second climax of the development of artificial neural networks. In addition to the BP algorithm, a variety of neural networks, such as SOM, ART, RBF, CC, RNN and CNN, also developed rapidly in this period.

i . Multi-layer neural network learning model

Multi-layer perceptron is an artificial neural network with forward structure, which is composed of input layer, hidden layer and output layer. The model has multiple node layers, each layer fully connected to the next layer, and the output is calculated one by one according to the hierarchical relationship. Starting from the outermost layer, the calculation of the deeper layer will be continued after the values of all units have been evaluated. Only after the values of all units of the current layer have been calculated, will the next layer be counted. This process is called forward propagation.

In a multi-layer neural network, each layer has a deeper abstract representation of the previous layer as the number of layers of the network increases, and each layer of neurons learns a more abstract representation of the values of the previous layer of neurons. For example, the first hidden layer learns the feature of edge, the second hidden layer learns the feature of shape composed of edge, the third hidden layer learns the characteristics of the pattern composed of shapes, and the last hidden layer learns the characteristics of the target composed of patterns.

The multi-layer neural network learning model overcomes the weakness that the perceptron cannot recognize linear inseparable data, and it has the ability to solve the complex problems quickly.

ii. BP neural network algorithm

Back propagation (BP) algorithm is an algorithm used in conjunction with optimization methods, such as gradient descent, and it is also a common method used to train artificial neural networks.

The BP neural network algorithm uses the gradient descent method to calculate and adjust the weight value in the negative gradient direction of the mean square error of a single sample[64]. The algorithm uses back propagation to transmit the error to the hidden layer of the neural network. The connection weight value between the output layer and the hidden layer is adjusted according to the error, and the threshold is determined in this process. The mean square error of the hidden layer neurons is calculated, and the connection weight and threshold between the input layer and the hidden layer are adjusted according to the error.

BP algorithm is mainly composed of two stages:

(1) Incentive communication: The propagation link in each iteration consists of two steps. The first step, forward propagation, sends the training input to the network to obtain an incentive response. The second step, back-propagation stage, calculates the difference between the excitation response and the target output corresponding to the training input, so as to obtain the response error of the output layer and the hidden layer.

(2) Weight update: The input value and response error are multiplied to obtain the gradient of the weight. The weight ratio will affect the speed and the effect of

training process, so it is called training factor. The direction of the gradient indicates the direction of error enlargement, so it needs to be reversed when updating the weight to reduce the error caused by the weight.

The two stages iterate repeatedly until the network's response to the input reaches a satisfactory predetermined target range.

iii. Self-organizing mapping network

Self-organizing mapping (SOM) is an artificial neural network that uses unsupervised learning to generate a low-dimensional discretized representation of the input space of training samples. The difference between SOM and other artificial neural networks is that it uses a proximity function to maintain the topological properties of the input space. The main goal of SOM is to convert the input signal of any dimension into one-dimensional or two-dimensional discrete map, and to perform this transformation adaptively in a topologically ordered manner.

The goal of learning in SOM is to make the different parts of the network have similar responses to the input pattern. The weight initialization of the neuron can adopt the minimal random value method or uniformly sample from the range of the two largest feature vectors. In the latter, the learning rate can be increased since the initial weight is a good approximation of the SOM weight.

The process of self-organization includes the following four aspects:

(1) Initialization: All connection weights are initialized with small random values.

(2) Competition: For each input mode, neurons calculate their respective discriminant function values to provide the basis for competition. The specific neuron with the smallest discriminant function value is declared the winner.

(3) Cooperation: The winning neuron determines the spatial position of the topological neighborhood of the excited neuron, thereby providing the basis for the cooperation between neighboring neurons.

(4) Adaptation: The value of the discriminant function associated with the input mode is reduced by adjusting the relevant connection weight appropriately, and the response of the winning neuron to the subsequent application of the similar input mode is enhanced.

iv. ART competitive learning network

Adaptive resonance theory (ART) neural network is a representative of competitive learning. Competitive learning is an unsupervised learning strategy. When this strategy is used, the output neurons of the network compete with each other, and only one neuron that wins the competition is activated at each moment.

ART alleviates the problem of plasticity and stability in competitive learning, in which plasticity means that the neural network must be able to learn new knowledge, and stability means that the neural network must maintain the memory of the previously learned knowledge while learning new knowledge, and it has strong environmental adaptability.

In practice, the recognition threshold has an important effect on the performance of the ART network. When the recognition threshold is high, the input samples will be divided into more and more detailed pattern classes. When the recognition is lower, there will be fewer and more rough pattern classes.

v. Radial basis function network

In 1985, Powell proposed a radial basis function (RBF) method for multivariate interpolation. RBF is a real value function whose value depends only on the distance from the origin. RBF neural network is developed from radial basis function, including input layer, hidden layer and output layer. The transformation from the input to the hidden layer is non-linear, and the transformation from the hidden layer to the output layer is linear.

The basic idea of the RBF network is to use RBF as the base of the hidden unit to form the hidden layer space. The input vector is directly mapped to the hidden space without weight connection, and the mapping relationship can be determined after the center point of RBF is determined.

In the RBF network, the mapping of the network from input to output is non-linear, and the network output is linear for the adjustable parameters, so the weight of the network can be solved directly by the linear equations. What is linearly indivisible in the low dimension becomes linearly separable in the high dimension, which greatly accelerates the learning speed and avoids local minimal problems.

vi. Recurrent neural network

Recurrent neural network (RNN) is a type of neural network. It takes sequence data as input. Recursion is performed in the evolutionary direction of the sequence, and all nodes are connected in a chain.

Researches on the recurrent neural network were began in the 1980s, which developed into one of the deep learning algorithms in the early 21st century[65]. Bidirectional RNN (Bi-RNN) and long short term memory (LSTM) networks are common recurrent neural networks.

The recurrent neural network has memory, parameter sharing and Turing completeness. It has certain advantages when learning the nonlinear characteristics of the sequence. Recurrent neural network is widely used in the natural language processing (NLP), such as speech recognition, language modeling, machine translation and other fields. It is also used in various time series forecasting problems. The recurrent neural network in which the convolutional neural network (CNN) is introduced can deal with computer vision problems involving sequential inputs[66-67].

2.1.5   Diversified development period

Through the combing of the first four stages, it can be seen that, although there are obvious distinguishing signs in each stage, almost all of them are developed around the evolution of artificial neural network methods and their learning rules. In fact, in addition to artificial neural networks, other algorithms in machine learning also emerged during these periods, such as the Boosting algorithm proposed by Schapire (1990), the AdaBoost algorithm proposed by Freund and Schapire (1995), the Bagging algorithm proposed by Breiman (1996) and the random forest algorithm proposed by Breiman (2001). In 2006, Hinton proposed deep learning network, the core idea is to solve the initial value selection problem of multi-hidden-layer neural networks through layer-by-layer learning, so as to improve the effect of classification learning. At present, integrated learning and deep learning have become the most popular research areas in machine learning.

The latest phase of machine learning began in 1986. On the one hand, due to the resurgence of neural network research, the study of the learning method of connectionism is in the ascendency, and the study of machine learning has come to a new climax in the whole world, and the study of the basic theory and integrated system of machine learning has been strengthened and developed. On the other hand, experimental research and applied research have received unprecedented attention. The rapid development of artificial intelligence technology and computer technology provides a new and more powerful research means for machine learning. Specifically, the symbolic learning changed from ignorant learning to growth-oriented learning, so the analytical learning with a certain knowledge background appeared during this period. Due to the progress of hidden nodes and back propagation algorithm, the neural network has made the connection mechanism learning algorithm come back and challenged the traditional learning[68]. At the same time, the evolutionary learning system and the genetic algorithm based on biological evolution have received much attention for absorbing the advantages of inductive learning and connection mechanism learning.

In 1989, Watkins proposed Q-learning model, which promoted in-depth study of reinforcement learning. The reinforcement learning system based on behaviorism shows new vitality because of the development of new algorithms and the new achievements in the application of genetic algorithms[69].

The important manifestations of machine learning entering a new stage are as follows:

(1) Machine learning has become an interdisciplinary subject and has formed a course in universities. It combines with applied psychology, biology, neurophysiology, mathematics, automation and computer science to form the theoretical basis of machine learning.

(2) The research of integrated learning system, which combines with various learning methods and makes up for each other's weaknesses, is emerging.

(3) The idea of the unity of machine learning and the fundamental problems of artificial intelligence is forming, such as the combination of learning and problem solving.

(4) The application scope of various learning methods is expanding, and some of them have become commodities. Knowledge acquisition tools of inductive learning have been widely used in diagnostic classification expert systems.

(5) Researches on data mining and knowledge discovery have formed an upsurge, and they have been successfully applied in biomedicine, financial management, commercial sales and other fields, which have injected new vitality to machine learning.

(6) Academic activities related to machine learning are unprecedentedly active. In addition to the annual machine learning seminar, there are also computer learning theory conferences and genetic algorithm conferences in the world.

i . Decision tree algorithm

In 1986, Australian computer scientist Ross Quinlan published the famous ID3 algorithm in *Machine Learning*, which led to the research of decision tree algorithms in machine learning.

Decision tree is an algorithm for decision making based on tree structure, and each single judgment evaluating the attributes of the feature in the decision-making process and generating the final feature judgment results based on the evaluation value[70]. The decision tree is composed of the following four parts: the root nodes composed of the complete set of samples, the internal nodes composed of feature attributes and leaf nodes, the leaf nodes composed of feature categories and branches, and the result output.

Decision tree is often used in the process of feature selection. Tree nodes are used to classify attributes and combine them into feature subsets. In the learning process, feature selection is completed in an auxiliary form.

The key point of decision tree learning is how to divide data characteristic attributes effectively. Different attribute partitioning methods will lead to different branch structures, which will affect the performance of the entire decision tree. The goal of attribute division is to divide each sub-node as effectively as possible, that is, the sub-nodes that are divided belong to the same category.

The decision tree is constructed by recursion method. The choice of leaf nodes is diverse, which can be used as the basis for leaf nodes selection when the node contains samples belonging to the same category, or the node attribute set is empty, or the attributes of all samples cannot be divided, or the sample set contained in the node is empty and cannot be divided, and it is set as the category with the most samples in the parent node.

In the construction of the decision tree, any complex classification problem can be identified by adding leaf nodes. Under the evaluation of single classification index, overreliance on training samples will lead to overfitting problems, which makes the effect of well-trained models on the test set unsatisfactory. Therefore, it is necessary to adopt two strategies of pre-pruning and post-pruning to reduce the overfitting problem of the decision tree algorithm.

In the decision tree construction process, the pre-pruning process removes many branches of the decision tree in the early stage, which reduces the training time, releases computing resources, and reduces the risk of overfitting. Pre-pruning also eliminates the subsequent sub-node branches of the current node, which reduces the branch expansion of the decision tree to a certain extent, and easily changes from overfitting to underfitting. On the one hand, post-pruning can retain more branches, and its performance is improved compared with pre-pruning. On the other hand, post-pruning requires traversing all nodes and calculating performance, and the training time is much longer than pre-pruning.

The workflow of the decision tree is as follows:

(1) Enter the data set to be classified and category label and target label.

(2) Check whether the data set has only one column, or whether the last

column has only one level.

(3) Call the information gain formula, calculate the information gain of all nodes, and obtain the category label corresponding to the maximum information gain.

(4) Establish a decision tree dictionary to save the current leaf nodes information.

(5) Enter the loop.

(6) Return to the decision tree dictionary.

(7) Calculate the sub-data sets in sequence according to the different levels of the category labels.

(8) Repeat steps (1) to (6) for the sub-data set to iterate.

ii. Random forest algorithm

In machine learning, random forest is a classifier containing multiple decision trees, and the output category is determined by the mode of the category output. The random forest algorithm is developed by Leo Breiman and Adele Cutler, and at the same time, Random Forests is their trademark. The term was derived from the random decision forests proposed by Tin Kam Ho of Bell Labs in 1995.

The characteristics of the random forest algorithm are as follows:

(1) A classifier with high accuracy can be generated for a variety of data.

(2) A large number of input variables can be processed.

(3) The importance of variables can be evaluated while determining the category.

(4) When constructing the forest, an unbiased estimate of the generalized error can be generated internally.

(5) Missing data can be estimated with accuracy even when a large portion of the data is missing.

(6) Variable interactions can be detected.

(7) Errors can be balanced.

(8) The closeness in each column can be calculated, which is useful for data mining, outlier detection and data visualization.

(9) The learning process is rapid.

(10) It can be extended and applied to unlabeled materials, that is, unsupervised gathering.

iii. Integrated learning model

In the motion of a cascade system, the features include the environmental information, such as personnel, vehicles, aircraft, tower, buildings, motor speed and steering angle. The features are complex and varied, and the learning of each feature will bring some problems such as large computation, overfitting and repeated operation. The integrated learning model is introduced to combine some features of the same category, and the group strategy learning of the combined features can improve the efficiency and accuracy.

Integrated learning generates a group of individual learners for the data of complex categories, and combines them with a variety of algorithm strategies. The model is shown in Figure 2 – 2.

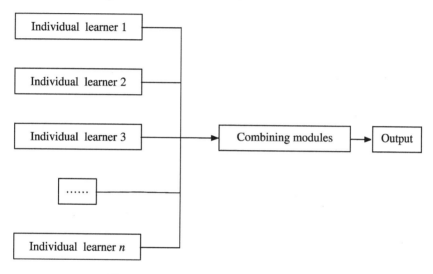

Fig. 2 – 2   Integrated learning architecture

Integrated learning includes homogeneous integration and heterogeneous integration. The individual learners in homogeneous integration are all in the same category, such as decision tree or neural network integration, and it is calculated by the base learner and base learning algorithm. The individual learners in the heterogeneous integration can combine multiple types of learning algorithms, which consist of component learners and component learning algorithms, such as decision tree and neural network. In integrated learning, it is evaluated with accuracy and diversity, the accuracy refers to the accuracy of individual learner classification and the diversity refers to the output difference between individual learners.

iv. Deep learning

The deep learning algorithm is extremely complex and powerful[71-72]. The more parameters are set, the more complex the model will be. More and more complex learning tasks can be completed. The commonly used deep learning architectures, such as DBN and CNN, are a way of feature learning representation. Multiple hidden layers are used to transform the initial input into the output target. The multi-layer processing method in deep learning can upgrade the low-level features to the high-level for characterization. Complex learning tasks can be realized by building simple models, and computations that cannot be completed by a single layer of mapping can be handled.

ⅴ. Reinforcement learning and inverse reinforcement learning

The reinforcement learning is an important branch of machine learning. Reinforcement learning involves placing the problem or object to be analyzed in the environment and giving certain incentives or punishments[73-74]. Through a large number of training, the network gradually produces the expected processing methods for the artificial rewards or punishments, so as to generate the maximum benefits of habitual response behavior.

The method of reinforcement learning is widely used in many fields. The strategy iteration is the most commonly used method in the reinforcement learning[75]. The purpose of strategy iteration is to make the strategy and the function value converge to the optimal value at the same time, which includes two sub-processes, strategy estimation and strategy update.

Unlike the reinforcement learning, the inverse reinforcement learning is suitable for solving the optimal strategy problem of unknown reward function[76-77]. The goal of the inverse reinforcement learning is to generate a potential return function structure based on observations and inferences from environmental models according to the parameterized characteristics of the predetermined reward function. The traditional inverse reinforcement learning method has established a relatively complete convergence theory; however, there are still shortcomings such as low learning efficiency and difficulty to evaluate the quality of the return function. The method of inverse reinforcement learning can not deal with the problems of the dimensionality disaster and combinatorial explosion in the continuous and high-dimensional Markov decision, so further study is needed to deal with this kind of problems.

ⅵ. Support vector machine algorithm

Support vector machine (SVM) is a kind of generalized linear classifier that classifies data in a supervised learning manner, and its decision boundary is the maximum margin hyperplane for solving learning samples.

SVM was proposed in 1964. It has been rapidly developed after the 1990s and has derived a series of improved and extended algorithms, which have been applied in pattern recognition such as portrait recognition and text classification. Since statistician Vapnik published SVM in *Machine Learning* in 1995, statistical learning represented by SVM had shined in the 1990s, and it quickly challenged the dominance of symbolic learning. At the same time, the introduction of integrated learning and deep learning has become an important extension of machine learning.

SVM uses hinge loss function to calculate empirical risk and adds regularization terms to the solution system to optimize structural risk. It is a sparse and robust classifier. SVM can perform non-linear classification through the kernel method, which is one of the common kernel learning methods.

SVM uses the principle of structural risk minimization to generate data dimensions[78-79]. SVM does not require the designer to provide upper-level knowledge for the input of the overall prediction structure. The computational complexity of the SVM has little relation with the original data dimension. It is suitable for the analysis of small and medium data samples and has global optimality.

2. 2 Feature engineering

Feature engineering is the process of transforming original data into features that express the essence of the problem better, and these features can be applied to the prediction model, which can improve the prediction accuracy of the invisible or incomprehensible data models.

The data features have a direct impact on the predicted model and the results. Generally, the better the feature selection and preparation, the better the results will be obtained. But the prediction results depend on many related attributes, such as the available data, prepared features, and model selection. Therefore, different projects need to adopt different feature selection and extraction methods.

The framework of the feature engineering knowledge system is shown in Figure 2 - 3.

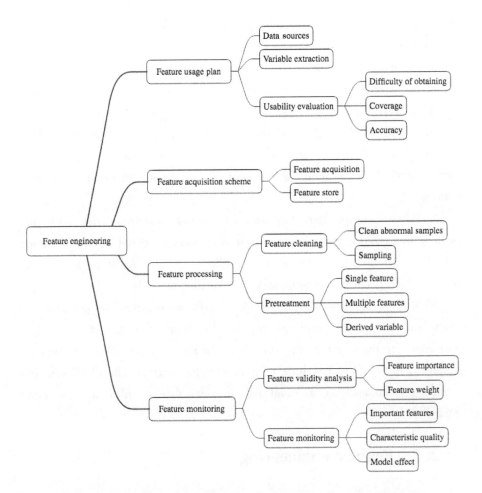

Fig. 2 – 3   Framework of the feature engineering knowledge system

2.2.1   Data structure

In machine learning, data structure is a way of storing and organizing data in a computer. It is a collection of data elements with structural characteristics. It studies the logical and physical structure of the data and the relationship between them. The appropriate operation is defined for this structure, and the corresponding algorithm is designed to ensure that the new structure obtained from these operations still retains the type of the original structure. That is, data structure is a collection of data elements of one or more specific relationships, which can be divided into logical structure and storage structure.

Data structure mainly includes scalar data structure, vector data structure, matrix data structure, and tensor data structure.

i . Scalar data structure

In physics, scalar refers to a physical quantity that remains unchanged under coordinate transformation, that is, a quantity with only magnitude and no direction. The calculations of these quantities follow general algebraic laws, such as mass, density, temperature, work, energy, distance, speed, volume, time, heat, resistance, power, potential energy, gravitational potential energy, electric potential energy and other physical quantities. No matter what coordinate system is selected, the value of the scalar will always remain unchanged.

Some scalars use positive and negative to indicate the magnitude, such as gravitational potential energy, electric potential, and electric charge. Positive charge means that the object is positively charged, while negative charge means that the object is negatively charged. Some scalars use positive and negative to indicate the trend, such as work. The positive and negative of work indicate the tendency of energy conversion. When the force does positive work on the object, the kinetic energy of the object increases. If the force does negative work on the object, the kinetic energy of the object decreases. The positive or negative of a scalar only represents the size and has nothing to do with the direction.

ii . Vector data structure

A vector refers to a quantity with size and direction, which can be visually represented as a line segment with an arrow. The direction pointed by the arrow represents the direction of the vector, the length of the line segment represents the size of the vector. Many physical quantities are vectors, such as the displacement of an object, the force by a ball hitting a wall.

The vector data structure expresses the object geometric position by recording the coordinates and the space relationship of the object. The vector data structure is to organize the data of the vector data model. Information of points, lines, and polygons can be represented as accurately as possible by recording coordinates and their relationships. The vector data structure records the coordinates of sampling points on the basis of geometric space coordinates, and its coordinate space is continuous.

iii. Matrix data structure

The concept of matrix was first proposed by the British mathematician Kelly in the 19th century. A matrix is a set of complex or real numbers arranged in a rectangular array. It first originated from a square matrix formed by the coefficients and constants of equations.

Matrix is a common tool in advanced algebra, and it is also common in applied mathematics, such as statistical analysis. In physics, matrices are used in circuits, mechanics, optics and quantum physics. A matrix is also needed for 3D animation production in computer science. The calculation of matrices is an important issue in the field of numerical analysis. It can simplify the calculation of matrices in theory and practical applications to decompose a matrix into a combination of simple matrices. For some matrices with a wide range of applications and special forms, such as sparse matrices and quasi-diagonal matrices, there are specific fast calculation algorithms, which are widely used in the fields of astrophysics and quantum mechanics.

In practical applications, there are matrices with very high order. If many elements are of the same value in the matrix, and their distribution has a certain law, which is called special matrix. Through special compilation and storage of these matrices, storage space can be greatly saved. The methods of compilation and storage are as follows:

(*i*) Symmetric matrix

A symmetric matrix is a matrix that is symmetric along the main diagonal. Only the lower triangle, including the main diagonal, can be stored. It is necessary to store $n \times n$ units, but now only $n \times (n+1)/2$ storage units are needed, which saves about half of the storage units. When n is large, not only a considerable amount of storage units can be saved, but also the amount of calculation can be greatly reduced in subsequent processing.

(*ii*) Sparse matrix

In a matrix, if the number of elements with a value of zero is far more than the number of non-zero elements and the distribution of non-zero elements is irregular, the matrix is called a sparse matrix. The method of triple storage is often used in sparse matrices. It is to store the row and column subscripts corresponding to the element while storing non-zero elements. Each non-zero element in the sparse matrix is uniquely determined by a triple, and all non-zero elements in the matrix are stored in an array composed of triples.

(*iii*) Array data structure

An array is a collection used to store multiple data of the same type. In order to facilitate the processing in program design, several elements of the same type are often organized in an orderly form, these ordered collections of similar data elements are called arrays.

An array is an ordered sequence of elements, and it is also a linear table data structure. The various variables that make up the array are called the component of the array or the element of the array. The numbers used to distinguish the elements of the array are called subscripts or subscript variables.

(*iv*) Tensor data structure

A tensor is a multilinear function that can be used to express the linear relationship between some vectors, scalars, and other tensors. The term tensor originated from mechanics. It was originally used to represent the stress state of each point in an elastic medium. Now, tensor theory is developed into a powerful mathematical tool for mechanics and physics.

A tensor is an extension of the concept of a vector, which is a multidimensional array that can be either a vector or a matrix, depending on the number of indexes. The first-order tensor can be expressed as an array, and the second-order tensor can be expressed as a matrix, and the third-order and above are called higher-order tensors.

A three-dimensional tensor is an array arranged in a grid according to a certain rule, and one of the variable numbers represents the axis, as shown in Figure 2 − 4. A tensor has three indexes, where the first index represents the row, and the second index represents the column, and the third index represents the axis. For example, V_232 points to the element in the second row, third column, and second axis.

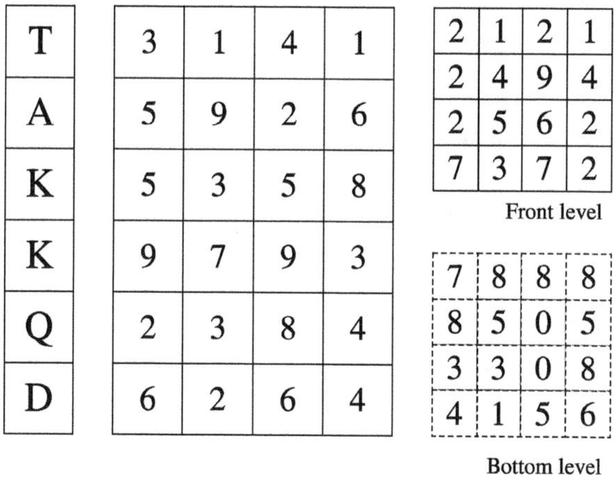

T	3	1	4	1
A	5	9	2	6
K	5	3	5	8
K	9	7	9	3
Q	2	3	8	4
D	6	2	6	4

2	1	2	1
2	4	9	4
2	5	6	2
7	3	7	2

Front level

7	8	8	8
8	5	0	5
3	3	0	8
4	1	5	6

Bottom level

Fig. 2 − 4　A three-dimensional tensor

2.2.2 Data loading

i . Load file

File loading includes three steps: defining data files, obtaining file objects, and reading file contents.

(*i*) Defining data files

Defining the data file is to assign the data file an object in advance. The purpose is to make subsequent operations more convenient and reduce code redundancy.

(*ii*) Obtaining file objects

The meaning of obtaining a file object is to generate an object based on the data file, and all subsequent operations on the data file are generated based on the object.

(*iii*) Reading file contents

Reading file contents refers to read the data in the file. In most cases, this is done by the function of READ or READLINES.

ii . Using the library to read data files

Third-party libraries are often used to read Excel, CSV, XML and other data files.

(*i*) Use Pandas library to read data

Pandas library can read data in a number of ways, the READ_CSV function is often used. The Pandas library requires a file path to be inserted into the data code. If the file to be read is saved in the Python working directory, there is no need to add the path, just use "filename. format".

(*ii*) Use Numpy library to read data

Numpy library can read data in three methods: LOADTXT, LOAD, and FROMFILE. The most commonly used one is LOADTXT. It allows users to set parameters and it is often used to process special data sets.

2.2.3 Data processing

Data processing includes data initialization, data standardization, data normalization and data nondimensionalization. Initialization is to analyze the weights of the convolutional neural network, including random initialization and all-zero initialization. Random initialization is to generate the weight value randomly, and all-zero initialization is to initialize the weight value to zero.

The core of quantitative feature binarization is to set a threshold value. If the value is greater than the threshold, it is assigned to one, and if the value is less than or equal to the threshold, it is assigned to zero, as shown in Equation (2 – 1).

$$x' = \begin{cases} 1, x > threshold \\ 0, x \leqslant threshold \end{cases} \qquad (2-1)$$

Nondimensionalization enables data of different specifications to be converted to the same specification. Nnondimensionalized methods include standardization and interval scaling. The premise of standardization is that the eigenvalues obey the normal distribution. After standardization, the data will be converted into a standard normal distribution. The interval scaling method uses boundary value information to scale the value interval of a feature to the range of a certain feature, such as [0,1]. Standardization needs to calculate the mean and standard deviation of the feature, as shown in Equation (2 – 2).

$$x' = \frac{x - \bar{x}}{s} \qquad (2-2)$$

Standardization is to process data according to the columns of the feature matrix, which converts the eigenvalues of the sample to the same dimension by calculating the z-score. Normalization is to process data according to the rows of the feature matrix. The purpose is to have a uniform standard for the sample vector, that is, to be transformed into a unit vector. The normalization formula is shown in Equation (2 – 3).

$$x' = \frac{x}{\sqrt{\sum_{j}^{m} x[j]^2}} \qquad (2-3)$$

There are many methods for the interval scaling. The common one is to use two maximum values for scaling, as shown in Equation (2 – 4).

$$x' = \frac{x - Min}{Max - Min} \qquad (2-4)$$

2.2.4 Feature processing

i . Feature types

The feature types include numerical type, category type, time type, text type, statistical type and combinatorial type.

(*i*) Numerical type

The numerical type indicates the size of the feature, and the signed value can represent the direction of the feature.

(*ii*) Category type

The category type is textual information, for example, the color is red, yellow or blue, and it needs to be processed firstly when the data is being stored.

(*iii*) Time type

The analysis of time features almost runs through every step of the entire feature engineering. Time features can be regarded as continuous values or discrete values. Time is often used as an important feature in data mining.

(*iv*) Text type

Text type refers to large paragraphs of text or multiple different documents.

(*v*) Statistical type

Statistical type includes addition and subtraction average type, quantile type, order type, and proportion type.

(*vi*) Combinatorial type

Combinatorial type includes splicing type and model feature combination type.

In actual engineering, the data obtained often have abnormal features. The methods based on statistics, distance and density can be used to distinguish the abnormal feature points.

The abnormal point detection algorithms based on statistics, such as range, interquartile, mean deviation, standard deviation, are suitable for univariate numerical data.

The abnormal point detection algorithms based on the distance mainly use the distance method to detect the abnormal points. A point with a distance greater than a certain threshold from most points in the data set is regarded as a abnormal point. The main distance measurement methods include absolute distance (Manhattan distance) , Euclidean distance and Mahalanobis distance.

The abnormal point detection algorithms based on density find local abnormal points by examining the density around the current point.

ii . Feature extraction

Feature extraction methods include forward search, backward search, and two-way search. The search process is shown in Table 2 – 1.

Table 2 – 1   Algorithm strategy and calculation process

Algorithm strategy	Calculation process
Forward search	Each feature is regarded as a feature subset to be recognized. Multiple optimal feature subsets are selected from all candidate subsets, and then the number of new features is gradually increased and the number of feature subsets is updated until the feature subset completes the search.
Backward search	All features are regarded as candidate feature subsets, and the optimal feature subset is identified gradually by removing the current optimal subset of feature subsets.
Two-way search	Combined with forward search and backward search, there are multiple operations of adding and kicking out feature subsets in each iteration.

iii. Feature selection

Feature selection is to select some of the most useful features that are most useful for result prediction from a number of features, and they are used as training features for machine learning. In the large-scale machine learning network, many features that are similar to each other or have an integrated effect on a certain result are often grouped into a feature subset, so as to reduce the amount of calculation and improve the accuracy of analysis.

Feature subset can be a set of control representations obtained from the analysis of environmental information such as traction speed, steering angle, accompanying personnel, vehicles, other aircraft, towers and buildings. This kind of feature is easy to recognize. The control of the tractor is based on the recognition of the subset feature. Feature selection can eliminate the attributes irrelevant to learning tasks and select the best feature subset. The way of iterating and generating a better candidate subset from the candidate feature subset can greatly reduce the complexity. For example, environmental features such as weather, temperature and wind speed have little influence on the control process, and they will not be included in the feature subset.

In the aircraft traction cascade system, the data set has discrete attributes. The preselected features can be divided into a number of subsets. For example, A1 is defined as {straight, curve, unrecognized}, and A2 is defined as {with navigation mark, no navigation mark, unrecognized}. The original data set can be divided into 3×3 subsets by matching A1 and A2 item by item. The selection process of the subset is roughly the same. The subset classification quality can be evaluated by calculating the information gain of the subset. The greater the gain, the better the effect in the classification, and the subset also contains more features that are helpful for classification, as shown in Equation(2 −5) and Equation(2 −6).

$$G(A) = \text{Ent}(D) - \sum_{v=1}^{V} \frac{|D^v|}{|D|} \text{Ent}(D^v) \qquad (2-5)$$

$$\text{Ent}(D) = - \sum_{i=1}^{|y|} p_{k\log_2 pk} \qquad (2-6)$$

Feature extraction is required for the identified features. The extraction methods include filter feature selection, wrapper-based feature selection, and embedded feature selection.

(*i*) Filter feature selection

The feature selection method that is separated from the learner is used in the filter feature selection, and manually selected features are used as a subset of the preset data for the training of the learner. When the identified feature is not in the preset range, it will be thrown out in the analysis to reduce the amount of system calculation. For example, the Relief algorithm uses correlated statistics to characterize feature attributes, each component of the statistics represents the importance of the corresponding feature, and the component of each statistic is calculated to obtain the most suitable feature subset.

In the cascade system, the characteristics affecting the motion such as wind speed, weather, temperature have low correlation with the motion process, and these features do not belong to the preset feature recognition range and do not have mutual relations, so they will be omitted in the filter feature selection. The characteristics such as motor speed and runway information are related to each other. The feature components of the cascade system are calculated to obtain the most appropriate feature subset.

The calculation of feature components is the core of filter feature selection. For each sample in the data set, the filter feature selection first finds the nearest neighbor of the sample. The neighbors of the same category are called the guessed neighbors, and the neighbors of the different categories are called the wrong guessed neighbors. Then each component of the statistic is calculated, as shown in Equation(2 −7).

$$\delta^j = \sum_i - diff(x_i^j, x_{i,nh}^j)^2 + diff(x_i^j, x_{i,nm}^j)^2 \qquad (2-7)$$

The distance of the J attribute is required to be as small as possible in the nearest guessing neighbors, and it is required to be as large as possible in the wrong guessing neighbors. The discrete attribute, Hamming distance, Manhattan distance and other parameters of the data sample are used to calculate the distance of each J attribute of the sample respectively, and the relevant statistics of the sample population can be obtained.

In order to enable the filter feature selection to be applied to the multiple classification problems, the Relief-F algorithm, a variant of the Relief algorithm, is used to re-describe the J component, as shown in Equation (2 – 8).

$$\delta^j = \sum_i - diff(x_i^j, x_{i,nh}^j)^2 + diff(x_i^j, x_{i,nm}^j)^2 + \sum_{l \neq k} pl + diff(x_i^j, x_{i,l,nm}^j)^2 \qquad (2-8)$$

Where,

pl represents the proportion of the first type sample in the data set.

In general, standard Relief has only one wrong guessed neighbor, while Relief-F has multiple wrong guessed neighbors.

(*ii*) Wrapper-based feature selection

Wrapper-based feature selection takes the selected learner as an evaluation criterion for feature selection. In each iteration of the calculation process, a new learner needs to be trained. The calculation is large, the complexity is high, and it takes up a lot of computing resources, but it has better system performance.

In the wrapper-based feature selection, the stopped parameter T can be set to automatically interrupt the calculation and release the computing capacity if there is no result after more than T times in the single calculation of searching for the optimal feature subset. Common algorithms and characteristics are shown in Table 2 – 2.

Table 2 – 2   Wrapper-based feature selection algorithms and features

Algorithm name	Features
Monte Carlo algorithm	The more sampling points and sampling times, the more approximate optimal solution can be obtained. The solution can certainly be obtained in the calculation process, but it is easy to get wrong solution.
Las Vegas algorithm	In the case of a large number of samples, it is easy to obtain the best solution, and the algorithm cannot guarantee convergence, but the solution is relatively correct and can be iterated continuously.

(*iii*) Embedded feature selection

Embedded feature selection is a process that integrates feature selection and learner training process. Both of them are put into the same optimization process, and the feature selection is automatically completed in the learner training process. In order to reduce the complexity of the model and improve the fit degree of the model, the L1 norm is often added to the loss function of the embedded feature selection, such as ridge regression and Lasso, as shown in Equation $(2-9)$ and Equation $(2-10)$.

$$\min_w \sum_{i=1}^{m} (y_i - w^T x_i)^2 + \lambda \parallel w \parallel_2^2 \qquad (2-9)$$

$$\min_w \sum_{i=1}^{m} (y_i - w^T x_i)^2 + \lambda \parallel w \parallel_1 \qquad (2-10)$$

2.2.5 Feature dimensionality reduction

When the feature selection is completed, the model can be trained. During the training process, because the feature matrix is too large, it leads to the problem of large amount of calculation and long training time. Therefore, it is essential to reduce the dimension of the feature matrix.

Dictionary learning and sparse coding are the common ways of feature dimensionality reduction. The algorithm uses the form of a matrix to map the original data into a sparse representation matrix through a dictionary. The sparse matrix fills the rows and columns of the matrix with a large number of zero elements, and all zero elements will not appear in the same row and column. In the environmental feature recognition of tractors, a large number of images, such as runway, road surface and sky, are filled with the same or similar pixels, as shown in Figure $2-5$. The areas of A, B, C, and D can be mapped to form a sparse matrix, which greatly reduces calculations.

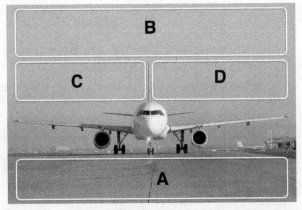

Fig. 2 –5   Areas with similar environmental characteristics

The sparse matrix is processed in a linearly separable manner, which can improve storage and learning efficiency, identify low-dimensional attributes and reduce noise. The calculation method is to solve the dictionary matrix B and the sparse representation α, as shown in Equation(2 – 11).

$$\min_{(B,\alpha_i)} \sum_{i=1}^{m} \| x_i - B\alpha_i \|_2^2 + \lambda \sum_{i=1}^{m} \| \alpha_i \|_1 \qquad (2-11)$$

The common dimensionality reduction methods include principal component analysis(PCA) and linear discriminant analysis(LDA). PCA and LDA have many similarities, and their essence is to map the original sample to a sample space with a lower dimensionality. But the mapping goals of PCA and LDA are different. PCA is to make the mapped samples have the maximum divergence, while LDA is to make the mapped samples have the best classification performance. So PCA is an unsupervised dimensionality reduction method, while LDA is a supervised dimensionality reduction method.

2.3   Learning methods for dealing with forecasting problems

2.3.1   Linear algebra and machine learning

The concepts in linear algebra are essential to understanding machine learning theory.

Linear algebra is a branch of mathematics that deals with linear relation problems. Linear relation means that the relation between mathematical objects is expressed in the form of a first order equation. For example, in analytic geometry, the equation of a straight line on a plane is a linear equation of two variables, the equation of the space plane is a three-dimensional linear equation, and a straight line in space is regarded as the intersection of two planes, which is represented by a set of equations composed of two ternary linear equations.

The research objects of linear algebra are vectors, vector spaces, linear transformations and finite-dimensional linear equations. Vector space is an important subject of modern mathematics, so linear algebra is widely used in abstract algebra and functional analysis. At present, the theory of linear algebra has been generalized to operator theory, and linear algebra can be expressed concretely through analytic geometry. Since nonlinear models in scientific research can be approximated as linear models, linear algebra is widely used in natural sciences and social sciences to model natural phenomena and perform efficient calculations.

2.3.2 Univariate linear regression

Linear regression is one of the simplest algorithms in machine learning, and it is an algorithm for supervised learning. The main idea of the algorithm is to learn a linear function on a given training set, solve the correlation coefficient under the constraint of the loss function, and finally test the regression effect of the model in the test set.

Linear regression is a statistical analysis method that uses regression analysis in data statistics to determine the quantitative relationship between two or more variables. The steps of linear regression analysis are as follows:

(1) Determine the independent variables and the dependent variables according to the prediction targets.

(2) Draw a scatter plot and determine the type of regression model.

(3) Estimate model parameters and establish the regression model.

(4) Test the regression model.

(5) Use the regression model to make predictions.

In linear regression analysis, only one independent variable and one dependent variable are included, and the relationship between the two variables can be approximated by a straight line. This kind of regression analysis is called unary linear regression analysis. If the regression analysis includes two or more independent variables, and the relationship between the dependent variable and the independent variable is linear, it is called multiple linear regression analysis. The significance test and error calculation are needed to determine whether the regression equation is valid and whether the estimated error meets the design requirements.

Univariate linear regression is also called unary linear regression. Univariate linear regression is a method of analyzing the linear correlation between only one independent variable. The value of a technical indicator is often affected by many factors. If only one factor is the main one and plays a decisive role, unary linear regression can be used for predictive analysis.

The predictive model of the unary linear regression is shown in Equation (2 - 12).

$$Y_i = ax_i + b \qquad (2-12)$$

Where,

x_i represents the value of the independent variable in period i.

Y_i represents the value of the dependent variable in period i.

a and b represent the parameters of the unary linear regression equation. a and b are obtained by the Equation $(2-13)$ and Equation $(2-14)$ respectively. n represents the number of samples.

$$a = \frac{n\sum x_i Y_i - \sum x_i \sum Y_i}{n\sum x_i^2 - (\sum x_i)^2} \qquad (2-13)$$

$$b = \frac{\sum Y_i}{n} - a\frac{\sum x_i}{n} \qquad (2-14)$$

The establishment steps of univariate linear regression model are as follows:

(1) Select the variables of the unary linear regression model.

(2) Draw calculation table and fitted scatter plot.

(3) Calculate the regression coefficients between variables and their correlation significance.

(4) Application of regression analysis results.

2.3.3   Multivariate linear regression

In regression analysis, if there are two independent variables, it is called binary regression; if there are more than two independent variables, it is called multiple regression. In the aviation industry, the occurrence of a phenomenon is often related to multiple factors. It is more effective and practical to predict or estimate the dependent variable by the optimal combination of multiple independent variables than by using only one independent variable. Therefore, multiple linear regression is of greater practical significance than unitary linear regression.

i . Binary regression model

(*i*) Model building

The binary regression model is shown in Equation $(2-15)$.

$$Y_i = a_0 + a_1 x_1 + a_2 x_2 + b \qquad (2-15)$$

(*ii*) Parameter estimation

The parameter estimation of binary regression is shown in Equation $(2-16)$ to Equation $(2-18)$.

$$\sum Y = na_0 + a_1 \sum x_1 + a_2 \sum x_2 \qquad (2-16)$$

$$\sum x_1 Y = a_0 \sum x_1 + a_1 \sum x_1^2 + b_2 \sum x_1 x_2 \qquad (2-17)$$

$$\sum x_2 Y = a_0 \sum x_2 + a_1 \sum x_1 x_2 + b_2 \sum x_2^2 \qquad (2-18)$$

(*iii*) Goodness of fit index

The calculation Standard error is shown in Equation (2 – 19).

$$SE = \sqrt{\frac{\sum (Y - Y')^2}{n - 3}} \qquad (2 - 19)$$

(*iv*) Confidence range

The confidence interval is shown in Equation (2 – 20).

$$\text{Confidence interval} = Y' \pm t_p SE \qquad (2 - 20)$$

Where,

t_p is the value in the t statistic table of the n or k degree of freedom; n is the number of observations; k is the number of variables including the dependent variables.

ii. Multiple linear regression

In the multivariate regression analysis, the matrix is often used for operation. The data set and weight vector are shown in Equation (2 – 21) and Equation (2 – 22) respectively.

$$D = \{ (x_1, y_1), (x_2, y_2), \cdots\cdots, (x_n, y_n) \} \qquad (2 - 21)$$

$$\theta (\theta_1, \theta_2, \cdots\cdots, \theta_n, \epsilon)^T \qquad (2 - 22)$$

The matrix X is composed of all eigenvectors and columns of weight 1, as shown in Equation (2 – 23).

$$X = \begin{bmatrix} x_1^1 x_2^1 \cdots x_m^1 1 \\ \cdots \ \cdots \ \cdots \\ x_1^n x_2^n \cdots x_m^n 1 \end{bmatrix} = \begin{bmatrix} x_1^T 1 \\ x_2^T 1 \\ \cdots \\ x_m^T 1 \end{bmatrix} \qquad (2 - 23)$$

2.4　Learning methods for dealing with classification problems

2.4.1　Logistic regression

Logistic regression is a generalized linear regression model, which has many similarities with multiple linear regression analysis. The two models are basically the same in model form, and the difference lies in their different dependent variables. In general, if the mapping function is a logistic function, it is a logistic regression. If the mapping function is a polynomial function, it is a polynomial regression. If it is continuous, it is a multiple linear regression. If it is a binomial distribution, it is a logistic regression. If it is Poisson distribution, it is Poisson regression. If it is a negative binomial distribution, it is a negative binomial regression.

The dependent variables of logistic regression can be binary or multiple. The binary logistic regression is widely used in engineering, and the softmax method is often used to deal with multi-class problems. Logistic regression is used in data mining, automatic diagnosis, economic forecasting and other fields. For example, the risk factors causing the fracture risk of aircraft structural parts are discussed, and the probability of fracture of structural parts is predicted according to the risk factors.

The applicable conditions of the logistic regression model are as follows:

(1) The dependent variable is binary, and it is a numerical variable.

(2) Residuals and dependent variables are binomial distributions.

(3) The independent variable has a linear relation with logistic probability.

(4) The observation objects are independent of each other.

2.4.2 Classification expression

i . Binary classification

Binary classification is often used in machine learning. Each sample is set to only one label, and the attribute of that label is represented by 0 or 1 in the binary classification. For example, whether the tire can still be used will be determined based on the picture of the airplane tire. In order to achieve this goal, all the training data should be divided into two categories, namely, the binary classification process, and then a classifier is trained. The image in the input training set is represented by the eigenvector, and the result of whether the tire is available is output. In the output results, 0 is used to indicate that the tire is unavailable, and 1 is used to indicate that the tire is available, so the engineering problem is transformed into a problem that the computer can handle.

ii . Multi-class classification

Multi-class classification means that there are multiple categories in the classification. For example, the aircraft in the pictures may be Boeing 737, Boeing 747, Boeing 787, Airbus A320, Airbus A321, Airbus A380, etc. Multi-class classification assumes that each sample has only one label. The picture of an airplane can be Boeing or Airbus, but it can't be both at the same time. In engineering, multi-class classification problems are often decomposed into many binary classification problems.

iii. Multi-label classification

Multi-label classification gives each sample a series of target labels, and it is believed that the attributes of the data samples are not mutually exclusive. For example, a topic related to aviation operations can be considered as an economic, political, financial, or aviation-related topic at the same time.

In the task of multi-label classification, the number of labels of a picture is not fixed. Some have one label, and some have two labels. It is easy to lose label data in the process of label encoding. In order to achieve the consistency and readability of the label information, the method of label complement is adopted; that is, the missing labels are all marked with 0, which converts the multi-label problem into a binary classification.

2.4.3 Logical distribution

Logistic distribution is a continuous probability distribution, including distribution function and probability density function. The distribution function is the integral of the probability density function, and the probability density function can be obtained by taking the derivative of the distribution function.

The distribution function and density function are shown in Equation(2 – 24) and Equation(2 – 25) respectively.

$$F(x) = P(X \leqslant x) = \frac{1}{1 + e^{-(x-\mu)/\gamma}} \qquad (2-24)$$

$$f(x) = F'(X \leqslant x) = P(X \leqslant x) = \frac{e^{-(x-\mu)/\gamma}}{\gamma(e^{-(x-\mu)/\gamma})^2} \qquad (2-25)$$

Where,

μ represents the position parameter;

γ is the shape parameter, $\gamma > 0$.

The logistic distribution is a continuous distribution defined by its location and scale parameters. The shape of the logistic distribution is similar to that of the normal distribution, but the tail of the logistic distribution is longer, so logistic distribution can be used to establish the data distribution with longer tail and higher wave peak than normal distribution.

2.4.4 Logistic regression solution

i. Stochastic gradient descent

The calculation of parameter update in the stochastic gradient descent(SGD) does not use a single sample, but a small number or a small batch of samples. SGD can reduce the variance in the parameter update process and make the convergence process more stable. At the same time, the calculation of cost and gradient can be vectorized, which is conducive to the use of depth optimization matrix algorithm in the calculation.

The learning rate in SGD is much smaller than the learning rate of batch gradient descent, because the former has greater variance during the update process. It is quite difficult to choose the right learning rate and learning rate change strategy. The actual effective standard method is to use a fixed learning rate small enough to provide stable convergence in one or two iterations at the beginning of the iteration, and the learning rate is halved as the convergence rate decreases. A further optimization method is to calculate the value of the objective function on the reserved data after each iteration, when the change value of the objective function in two adjacent iterations is less than a certain small threshold, the learning rate is reduced. This tends to converge nicely to a local optimum.

The SGD algorithm can converge well to a local optimal value. Whether the local optimal value is the global optimal value requires further judgment. In order to achieve better training results, the data is generally shuffled randomly before each iteration to obtain better training uncertainty.

The iterative update method of the SGD algorithm is shown in Equation (2 – 26) and Equation (2 – 27).

$$S_i = \frac{\partial J(w)}{\partial w_i} = (p(x_i) - y_i) x_i \qquad (2-26)$$

$$w_i^{k+1} = w_i^k - \alpha g i \qquad (2-27)$$

Where,

k is the number of iterations.

After each parameter update, the iteration is stopped by comparing whether the difference between the values of the loss function before and after the update is less than the threshold or reached the maximum number of iterations.

ii . Newton's method

The basic idea of Newton's method is to perform the second-order Taylor expansion of the function near the estimated value of the existing minimum point, and then find the next estimated value of the minimum point.

Assuming that w^k is the current minimum estimate, then,

$$\varphi(w) = J(w^k) + J'(w^k)(w - w^k) + \frac{1}{2} J''(w^k)(w - w^k)^2 \qquad (2-28)$$

If $\varphi'(w) = 0$, then,

$$w^{k+1} = w^k - \frac{J'(w^k)}{J''(w^k)} \qquad (2-29)$$

The iterative update formula is shown in Equation(2 – 30).

$$w^{k+1} = w^k - \frac{J'(w^k)}{J''(w^k)} = w^k - H^{-1}k \times gk \qquad (2-30)$$

Where,

$H^{-1}k$ is the Hessian matrix, and the Newton's method requires the objective function to be second-order continuously differentiable in the calculation process, as shown in Equation(2 – 31).

$$H_{mn} = \frac{\partial^2 J(w)}{\partial w_m \partial w_n} = h_w(x^{(i)})(1 - p_w(x^{(i)}))x_m^{(i)} x_m^{(i)} \qquad (2-31)$$

iii. Regularization

(*i*) L1 regularization

L1 regularization is equivalent to adding a priori knowledge to the model based on LASSO regression, as shown in Equation(2 – 32).

$$f(w|u,\sigma) = \frac{1}{2\sigma}\exp\left(-\frac{|w-u|}{\sigma}\right) \qquad (2-32)$$

Where,

w obeys a Laplace distribution with zero mean.

Due to the introduction of priori knowledge, the likelihood function is shown in Equation(2 – 33) and Equation(2 – 34).

$$L(w) = P(y|w,x)P(w) \qquad (2-33)$$

$$L(w) = \prod_{i=1}^{N} p(x_i)^{y_i}(1 - p(x_i))^{1-y_i} \prod_{j=1}^{d} \frac{1}{2\sigma}\exp\left(-\frac{|w_j|}{\sigma}\right) \qquad (2-34)$$

The objective function is shown in Equation(2 – 35).

$$-\ln L(w) = -\sum_i [y_i \ln p(x_i) + (1-y_i)\ln(1-p(x_i))] + \frac{1}{2\sigma^2}\sum_j |w_j| \qquad (2-35)$$

(*ii*) L2 regularization

L2 regularization is equivalent to adding a prior knowledge to the model based on Ridge regression, as shown in Equation(2 – 36).

$$f(w|u,\sigma) = \frac{1}{\sqrt{2\pi}\sigma}\exp\left(-\frac{(w-u)^2}{2\sigma^2}\right) \qquad (2-36)$$

Where,

w obeys a normal distribution with zero mean.

After introducing the prior knowledge, the likelihood function is shown in Equation(2-37).

$$L(w) = P(y|w,x)P(w) = \prod_{i=1}^{N}p(x_i)^{y_i}(1-p(x_i))^{1-y_i}\prod_{j=1}^{d}\frac{1}{\sqrt{2\pi}\sigma}$$

$$\exp(-\frac{w_j^2}{2\sigma^2}) = \prod_{i=1}^{N}p(x_i)^{y_i}(1-p(x_i))^{1-y_i}\frac{1}{\sqrt{2\pi}\sigma}\exp(-\frac{w^Tw}{2\sigma^2})$$

$$(2-37)$$

The objective function is shown in Equation(2-38).

$$-\ln L(w) = -\sum_i[y_i\ln p(x_i) + (1-y_i)\ln(1-p(x_i))] + \frac{1}{2\sigma^2}w^Tw$$

$$(2-38)$$

2.5 Neural network

Artificial neural network uses multiple neurons to build hidden layers of different structures, which can realize regression and classification of nonlinear features and fit any function under appropriate conditions and enough training samples. Artificial neural network does not filter samplesin training, it connects the weight coefficient of the network with each node. The artificial neural network can constantly fit the data model and minimize the risk under the input of upper-level experience and knowledge, which is suitable for model regression classification under specific environment, and also suitable for analysis of large data samples. By integrating other machine learning algorithms, the new generation of artificial neural networks, such as GRNN, can effectively prevent the local but not global optimal problems in the convergence process of artificial neural network, and it also has the advantages of fast convergence and high fitting accuracy of data samples[80-83].

Neural network is a mathematical algorithm that imitates the characteristics of neurons and processes distributed information in parallel. The network adjusts the connection relationships and weight values of internal nodes by itself to obtain a more accurate recognition model.

2.5.1 Machine learning models

Machine learning models include supervised learning, semi-supervised learning and unsupervised learning. The three learning methods are all based on analyzing and learning a large number of data samples and mapping them into corresponding rules to realize the automatic operation process without manual intervention. The difference among the three learning methods is whether the training samples contain the label and state information in the manual operation[84-85].

The goal of the three different supervised learning machine algorithms is to find a function that can best map a set of inputs to its correct output.

i . Supervised learning

Supervised learning refers to the process in which the parameters of each module of the machine learning network are first obtained through manual operation and annotation, so that the machine learning network can operate the system according to the existing knowledge and meet the performance requirements when encountering similar problems. Supervised learning must rely on the label information of the existing training data, these labeled information include training examples and they have parameter characteristics that can be input into the system network. Supervised learning generates a processing model for the problem by analyzing a large number of training data examples, each example consisting of an input object and an output value. When a new example appears, the example is used as an input object, and an output value is obtained through the supervised learning network. A good supervised learning network requires high-quality data and classification labels, and it is closely related to the loss function, data flow and other architectures.

Supervised learning extrapolates a machine learning task from the labeled training data. The training data includes a set of training examples. In supervised learning, each instance is composed of an input object (usually a vector) and a desired output value (also called supervised signal). The supervised learning algorithm analyzes the training data, and it generates an inference function which can be used to map the new examples. An optimal solution would allow the algorithm to correctly determine the class labels of those invisible instances. This requires the learning algorithm to be formed in a reasonable way from training data to invisible situations.

ii . Semi-supervised learning

Semi-supervised learning is a key issue in the field of pattern recognition and machine learning. It is a learning method combining supervised learning and unsupervised learning. Semi-supervised learning uses large amounts of unlabeled data, and labeled data at the same time, for pattern recognition.

There are three basic assumptions in semi-supervised learning to establish the relationship between the prediction sample and the learning goal. It includes smoothness assumption, cluster assumption and manifold assumption. In essence, these three kinds of assumptions are the same, but the focus of each one is different, in which the manifold assumption is more universal.

iii. Unsupervised learning

In engineering, due to the lack of sufficient prior knowledge, it is difficult to manually label categories, or the cost of manual category labeling is too high. Therefore, how to use computers to deal with such problems has become a hot research topic at present. It is called unsupervised learning to solve various problems in pattern recognition according to the unknown category of unlabeled training samples.

Unsupervised learning algorithms mainly include principal component analysis (PCA), isometric mapping, local linear embedding and Laplace feature mapping, etc.

In principle, PCA and other data dimensionality reduction algorithms are suitable for deep learning, but these data dimensionality reduction methods are more complex, and the target of the algorithm is too clear, so that there is no secondary information in the abstracted low-dimensional data, which may be the main factor to distinguish data from a higher level. Therefore, the unsupervised learning methods used in deep learning usually adopt relatively simple algorithms and intuitive evaluation criteria.

At present, in deep learning, unsupervised learning is mainly divided into two categories. One is the deterministic self-encoding method and its improved algorithm, and the main goal is to recover the original data as lossless as possible from the abstracted data. The other is the probabilistic restricted Boltzmann machine and its improved algorithm, and the main goal is to maximize the probability of the original data when the restricted Boltzmann machine reaches a stable state.

(*i*) Deterministic unsupervised learning

Deterministic unsupervised learning includes auto-encoding, sparse auto-encoding and denoising auto-encoding. Auto-encoding can be seen as a special three layer BP neural network; its particularity is reflected in the need to make the input and output of the auto-encoding network as similar as possible, that is, to make the encoding as lossless as possible. Although sparse auto-encoding can learn an equality function to make the visible layer data as equal as possible to the encoded and decoded data, but its robustness is still poor, especially when the probability distribution of the test samples and the training samples are quite different. For this reason, Vincent proposed denoising auto-encoding based on the sparse auto-encoding to improve the robustness of the algorithm.

(**ii**) Probabilistic unsupervised learning

The typical representative of probabilistic unsupervised learning is the restricted Boltzmann machine. The restricted Boltzmann machine is a simplified version of the Boltzmann machine. Through the method of probability statistics, the activation state of the hidden layer can be easily calculated from the visible layer data.

2.5.2 Loss function

Loss function is used to estimate the degree of inconsistency between the predicted value of the model and the true value. It is a non-negative and real-valued function. In the fields of optimization, statistics, econometrics, decision theory, machine learning and computational neuroscience, loss function or cost function refers to a function that maps an event to a real number that expresses the economic cost or opportunity cost associated with the event, thereby, the association of some costs with the event can be visually expressed. The goal of an optimization problem is to minimize the loss function by setting the objective function as the loss function or its negative value, so the engineering optimization problem can be transformed into a mathematical problem for solving the minimum value of the loss function.

Loss function can be divided into empirical risk loss function and structural risk loss function. The empirical risk loss function is the difference between the predicted result and the actual result, and the structural risk loss function is a function of the empirical risk loss function and the regular term (L_0, L_1, L_2).

Loss function is the core of the empirical risk function, and it is also an important part of the structural risk function. The risk structure of the model includes risk terms and regular terms, as shown in Equation(2 – 39).

$$\theta^* = \arg \min_\theta \frac{1}{N} \sum_{i=1}^{N} L(y_i, f(x_i, \theta) + \lambda \Phi(\theta)) \qquad (2-39)$$

Where,

L represents the loss function;

$\Phi(\theta)$ is the regularization term, the norm of L_1 or L_2 is often used.

The process of solving θ^* is the process of calculating the minimum value of the objective function, the mean value function represents the empirical risk function.

The most common regression loss functions are MAE, MSE, RMSE, Huber Loss, Log-Cosh, Quantile Loss, etc. The commonly used classification loss functions are Log loss, CE, KL divergence, logistic loss, Focal loss, Hinge loss, exponential loss, etc. Different loss functions should be used in different machine learning problems, and the main basis for selection is as follows:

(1) The algorithm of machine learning.

(2) Whether the selected function is easy to calculate the derivative.

(3) The proportion of outliers in the data set.

i . Regression loss function

(*i*) Absolute value loss function

The absolute value loss function is to calculate the absolute value of the difference between the predicted value and the target value, as shown in Equation (2 – 40).

$$L(Y, f(X)) = |Y - f(x)| \tag{2-40}$$

The most common absolute value loss functions include the mean absolute error loss function (MAE) , as shown in Equation (2 – 41) , and the mean absolute percentage error function(MAPE) , as shown in Equation(2 – 42).

$$\text{MAE}: L = \frac{1}{n} \sum_{i=1}^{n} |(y_{\text{true}}^{(i)} - y_{\text{pred}}^{(i)})| \tag{2-41}$$

$$\text{MAPE}: L = \frac{1}{n} \sum_{i=1}^{n} \frac{|(y_{\text{true}}^{(i)} - y_{\text{pred}}^{(i)})|}{y_{\text{true}}^{(i)}} \times 100 \tag{2-42}$$

(*ii*) Squared loss function

The standard form of the mean squared error(MSE) loss function is shown in Equation(2 – 43).

$$\text{MSE}: L = \frac{1}{n} (y_{\text{true}}^{(i)} - y_{\text{pred}}^{(i)})^2 \tag{2-43}$$

(*iii*) Mean absolute error

The standard form of *MAE* loss function is shown in Equation(2 – 44).

$$MAE = \frac{1}{N} \sum_{i=1}^{N} |y_i - f(x_i)| \tag{2-44}$$

MAE calculates only the average value of the error without considering the direction. All samples have the same weights for the residuals of the average, so *MAE* has better robustness to the outliers. But since the derivative of *MAE* is constant, even a small error will cause the gradient to increase, which is not conducive to the network convergence.

(**iv**)Smooth L_1 loss function

The standard form of the smooth L1 loss function is shown in Equation (2 - 45).

$$\text{Smooth } L_1 = \begin{cases} 0.5x^2 & |x| < 1 \\ |x| - 0.5 & |x| \geqslant 1 \end{cases} \qquad (2-45)$$

Smooth L_1 loss function is smoother and more stable than L_1 loss function, and it converges faster. Compared with the L_2 loss function, smooth L_1 is insensitive to the outliers, the gradient change is relatively small and the convergence is easier during training.

(**v**)Mean squared error loss function

The standard form of *MSE* loss function is shown in Equation(2 - 46).

$$MSE = \frac{1}{N} \sum_{i=1}^{N} (y_i - f(x_i))^2 \qquad (2-46)$$

Only the average size of the error, without its direction, is considered for the *MSE*. In the *MSE* calculation process, all points are continuous and smooth, and the gradient is easy to calculate, and it has a relatively stable solution. However, if the error is greater than one, the error will be further increased after being squared in the *MSE*, which will lead to the poor robustness[86]. And when the input value of the function is far from the center value, it may lead to the gradient explosion problems to use gradient descent method.

(**vi**)Root mean squared error loss function

The standard form of the root mean squared error (RMSE) loss function is shown in Equation(2 - 47).

$$RMSE = \sqrt{\frac{1}{N} \sum_{i=1}^{M} (y_i - f(x_i))^2} \qquad (2-47)$$

The calculation process of the RMSE is relatively intuitive, the dimension of *RMSE* is similar to the input data compared with *MSE*. *RMSE* is prone to the calculation problem of gradient explosion.

ii . Classification loss function

(**i**)Logarithmic loss function

The standard form of the logarithmic loss function is shown in Equation (2 - 48).

$$L(Y, P(Y|X)) = -\log P(Y|X) \qquad (2-48)$$

The mean squared logarithmic error(MSLE) is shown in Equation($2-49$).

$$\text{MSLE}: L = \frac{1}{n}\sum_{i=1}^{n}(\log(y_{true}^{(i)}+1)-\log(y_{pred}^{(i)}+1))^{2} \qquad (2-49)$$

Features of the logarithmic loss function is as follows: The logarithmic loss function is a very good representation of the probability distribution, and it is very suitable for the situations where the confidence level of the results in each category should be known. The robustness is not strong, and it is more sensitive to the noise than Hinge loss. The loss function of logistic regression is the logarithmic loss function.

(*ii*) Hinge loss function

Hinge loss function is mainly used in support vector machines. The name is derived from the graph of its loss function, which is a polyline. The general function expression is shown in Equation($2-50$).

$$L(m_{i}) = \max(0,1-m_{i}(w)) \qquad (2-50)$$

The loss function can be further written as shown in Equation($2-51$).

$$L = \frac{1}{n}\sum_{i=1}^{n}(\max(0,1-y_{true}^{(i)}\times y_{pred}^{(i)})) \qquad (2-51)$$

The common improvement is the hinge loss variance loss function, and the function is shown in Equation($2-52$).

$$L = \frac{1}{n}\sum_{i=1}^{n}(\max(0,1-y_{true}^{(i)}\times y_{pred}^{(i)}))^{2} \qquad (2-52)$$

Features of hinge loss function are as follows: The loss value of the hinge loss function is zero if it is classified correctly, otherwise the loss value is $1-yf(x)$. It is often used in SVM algorithm. In general, $f(x)$ is the predicted value, it is between -1 and $+1$, and y is the target value(-1 or 1), and $|f(x)|>1$ is discouraged. That is, the classifier is not encouraged to be overconfident, and there will be no reward if a correctly classified sample is more than one away from the dividing line, so that the classifier can focus on the overall error. The robustness is relatively high, and it is not sensitive to abnormal points and noise.

(*iii*) Zero-one loss function

The value of the zero-one loss function is one if the predicted value and the target value are not equal, otherwise it is zero, as shown in Equation($2-53$).

$$L(Y,f(X)) = \begin{cases} 1, Y \neq f(X) \\ 0, Y = f(X) \end{cases} \qquad (2-53)$$

Features of the zero-one loss function: The zero-one loss function corresponds to the number of incorrect classification. The function is used in the perceptron, but the condition of equality is too strict, so the condition can be relaxed, that is, it is considered equal when $|Y - f(x)| < T$ is satisfied, as shown in Equation $(2-54)$.

$$L(Y, f(X)) = \begin{cases} 1, |Y - f(X)| \geqslant T \\ 0, |Y = f(X)| < T \end{cases} \qquad (2-54)$$

(**iv**) Exponential loss function

The standard form of the exponential loss function is shown in Equation $(2-55)$.

$$L(Y|f(X)) = \exp[-yf(x)] \qquad (2-55)$$

The exponential loss function is very sensitive to the outliers and noise, and it is used in the AdaBoost algorithm.

(**v**) Perceptron loss function

The standard form of the perceptual loss function is shown in Equation $(2-56)$.

$$L(y, f(X)) = \max(0, -f(x)) \qquad (2-56)$$

Features of the perceptual loss function: Perceptual loss function is a variant of the hinge loss function. The hinge loss function penalizes the points near the judgment boundary very strongly, while the perceptron loss function continues to iterate as long as the judgment category of the sample is correct, regardless of the distance of the judgment boundary. The perceptual loss function is simpler than hinge loss function, but the generalization ability of the model is weaker than hinge loss function.

(**vi**) Cross-entropy loss function

The standard form of the cross entropy loss function is shown in Equation $(2-57)$.

$$C = -\frac{1}{n} \sum_{x} [y \ln a + (1-y) \ln(1-a)] \qquad (2-57)$$

Where,

x represents the sample;

y represents the actual label;

a represents the predicted output;

n represents the total number of samples.

Features of the cross entropy loss function: The cross-entropy loss function is essentially a log-likelihood function, which can be used in binary classification and multi-classification tasks. The cross-entropy loss function in the binary classification is shown in Equation(2 – 58), where the input data is the output of the softmax or sigmoid function.

$$\text{loss} = -\frac{1}{n}\sum_x y\ln a + (1 - y)\ln(1 - a) \qquad (2-58)$$

The cross-entropy loss function in the multi-classification problem is shown in Equation(2 – 59), where the input data is the output of the softmax or sigmoid function.

$$\text{loss} = -\frac{1}{n}\sum_i y_i \ln a_i \qquad (2-59)$$

When the sigmoid is used as the activation function, the cross entropy loss function is often used instead of the mean square error loss function, because it can solve the problem that the weight update of the square loss function is too slow. The cross-entropy loss function has the good properties that the weight update is fast when the error is large, and the weight update is slow when the error is small.

2.5.3   Convolutional neural network

Convolutional neural network(CNN) is a feedforward neural network consisting of one or more convolutional layers and fully connected layers. CNN also includes association weights and pooling layers[87]. This structure enables the convolutional neural network to use the two-dimensional structure of the input data to obtain better recognition results. Compared with other deep learning structures, convolutional neural networks can give better results in image and speech recognition. This model can also be trained using back CNN propagation algorithms. Compared with other feedforward neural networks, need to consider fewer parameters, which makes CNN an attractive structure for deep learning.

CNN is a calculation method that applies convolution calculation to feedforward network. It is composed of a data input layer, a convolutional layer of different convolution kernels, a pooling layer that matches the output, a fully connected layer which is connected to each neuron, and an output layer[88 – 91]. It is suitable for the problems of image classification and recognition.

i . Input layer

The input layer of CNN can handle multidimensional data. The input layer of one-dimensional CNN receives one-dimensional or two-dimensional arrays, where one-dimensional arrays are usually time or spectrum samples, and the two-dimensional arrays may contain multiple channels. The input layer of two-dimensional CNN receives two-dimensional or three-dimensional arrays. The input layer of three-dimensional convolutional neural network receives four-dimensional arrays.

CNN is widely used in the field of computer vision. Many studies presuppose three-dimensional input data when introducing its structure, that is, the two-dimensional pixels on the plane and the RGB channel. The gradient descent algorithm is used in the convolutional neural network, though its input features need to be standardized. This is similar to the other neural network algorithms. The input data should be normalized in the channel or in the time or frequency dimension before the learning data is input into the convolutional neural network. If the input data is image pixels, the original pixel value can be normalized to the range of zero to one. In the subsequent processing, the standardization of input features is beneficial to improve the learning efficiency and performance of the convolutional neural network.

ii . Convolutional layer

Convolutional layer is a set of parallel feature maps. It is composed by sliding different convolution kernels on the input image and running certain operations. The sliding process is called stride, and the stride is a factor that controls the size of the output feature map.

The size of the convolution kernel is much smaller than the input image, and it overlaps or acts in parallel to the input image. All elements in a feature map are calculated by the convolution kernel, that is, a feature map shares the same weight and the bias terms.

The convolutional layer is the most important layer in the convolutional neural network, and it is also the source of the name convolutional neural network. There are two key operations in the convolutional layer, one is the local correlation which treats each neuron as a filter for processing, the other is the receptive field, and the filter is used to calculate the local data, as shown in Figure 2 − 6.

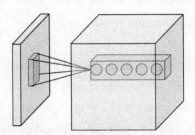

Fig. 2 − 6   Convolutional layer

iii. Active layer

After convolution, a bias term is usually added and a nonlinear activation function is introduced for processing. The bias is not related to the element position; it is only related to the layer. The most common activation functions are ReLU function, sigmoid function, tanh function, etc. The graphs are shown in Figure 2 – 7, Figure 2 – 8 and Figure 2 – 9 respectively.

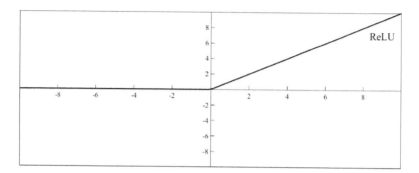

Fig. 2 – 7   ReLU function

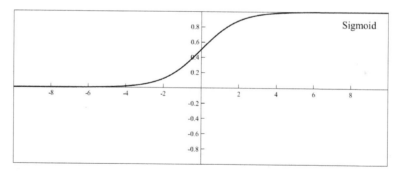

Fig. 2 – 8   Sigmoid function

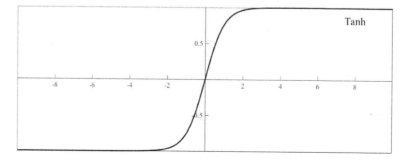

Fig. 2 – 9   Tanh function

The activation layer can enhance the nonlinear characteristics of the decision function and the entire neural network without changing the convolutional layer itself. In fact, some other functions can also be used to enhance the nonlinear characteristics of the network, such as the hyperbolic tangent function. Compared with other functions, ReLU function is more popular, because ReLU can increase the training speed of the neural network by several times without affecting the generalization accuracy of the model significantly.

The action process of the activation function is shown in Figure 2 – 10.

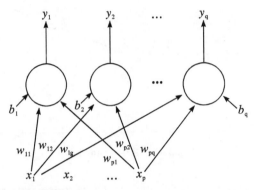

Fig. 2 – 10   The action process of the activation function

iv. Pooling layer

Pooling is another important concept in the convolutional neural network. From the working principle, pooling is a non-linear form of downsampling. In image processing, pooling can play a role in the feature dimensionality reduction. The information contained in an image is huge; at the same time, there are many features, and much of the information is duplicated and is not useful for the image processing task. The pooling layer can remove this kind of redundant information: the most important features can be extracted, and it prevents overfitting and is more convenient for the optimization to a certain extent.

The reason why the pooling layer can effectively improve the performance of the machine learning is that the precise position of a feature is far less important than its rough position relative to other features. The pooling layer will continuously reduce the size of the data space, so the number of parameters and the amount of calculation will also decrease, which also controls over-fitting to a certain extent. In general, the pooling layer is inserted periodically between the convolutional layers in the CNN network structure. Since the convolution kernel is a feature finder, various edges in the image can be easily found through the convolution layer. However, the features found by the convolutional layer are often too accurate. Even if an object is shot continuously at high speed, the edge pixel positions of the object in the photo are unlikely to be exactly the same. The sensitivity of the convolutional layer to the edge can also be reduced by the pooling layer.

There are many different forms of non-linear pooling functions in machine learning, among which the max pooling is the most common one. The max pooling divides the input image into several rectangular regions, and it outputs the maximum value for each subregion. The working mode is shown in Figure 2 – 11.

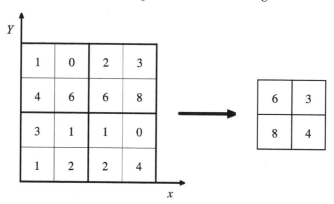

Fig. 2 – 11   Working mode of the max pooling

In addition to the max pooling, the pooling layer can also use other pooling functions, such as average pooling and L2-norm pooling. In the past, the average pooling was widely used, but now, it has become less commonly used due to the better performance of the max pooling in practice. Another type of Region of Interest (RoI) pooling is a variant of the max pooling, where the output size is fixed and the input rectangle is a parameter.

v. Fully connected layer

After several convolution and the max pooling layer, the advanced inference calculation process in the neural network is completed by the fully connected layer. As in the conventional non-convolutional artificial neural network, the neurons in the fully connected layer are related to all the activations in the previous layer, so the activation of the fully connected layer can be calculated as an affine transformation, that is, multiplying by a matrix and adding a bias(vector plus a fixed or learned bias) to obtain the overall perception of machine learning.

vi. Output layer

After several times of input, convolution, activation, pooling and full connection, the convolutional neural network is finally connected to the output layer. In the output layer, the model will generate corresponding results and provide accuracy on the test data according to the training data.

Before the fully connected layer, overfitting may occur if the number of neurons is too large and the learning ability is strong. In order to reduce the overfitting, the dropout operation can be appropriately introduced to delete part of neurons randomly in the neural network. At the same time, the robustness can be increased by local normalization and data enhancement.

2.6　Intelligent image processing methods

Image processing technology includes digital image processing and analog image processing. The digital image processing technology involves many disciplines, such as mathematics, computer science, information science, physics and biology[92-94]. It is suitable for handling complex nonlinear problems, and it has the characteristics of high processing accuracy and rich processing content. The analoy image processing technology is mainly used in geometric processing, image enhancement, image reconstruction, image coding, image restoration and image recognition[95], and it includes inter-frame difference method, optical flow method, image segmentation method, and background difference method[96].

The method of inter-frame difference analyzes the difference between the consecutive frames or multiple frames in the same video[97]. It compares the gray value of each pixel to achieve the target detection and extraction. The method has the characteristic of fast calculation speed, small calculation amount and easy implementation, and it is widely used in the video data real-time processing.

The method of optical flow is a detection segmentation algorithm based on the optical flow estimation[98]. It does not need to input the scene structure information, and it only calculates the pixel motion information between the adjacent video frame sequences, which greatly reduces the amount of the calculation, and reduces the problem of image classification and recognition depending on the background learning and background models.

The method of image segmentation recognizes the image gray scale, edge and texture features, and divides them into different objects and identified areas[99]. The identified region and target are detected according to the pre-set identification and matching method. The edge detection operators include the first-order and the second-order differential operator: the first-order operator is commonly used in Roberts, Prewitt, Sobel, Canny, etc. , and the Laplacian and LoG operators are commonly used in the second-order operator.

The method of background difference is used to calculate the difference between the input image and the background image, and the moving target is divided directly according to the image gray level[100]. The background image is changing all the time, so it is necessary to keep the corresponding update frequency of the background image in the continuous target detection process.

Chapter Three

Simulation Design of
Aircraft Traction Cascade System

The motion process of the aircraft traction cascade system contains many motion states, such as traction, pushing, warehousing, etc. , and their functions are complex. The motion process is constrained by the environment, so it is necessary to adjust the attitude of aircraft according to the motion state of surrounding vehicles and people at all times. The traction cascade system, which is composed of aircraft and tractor, is shown in Figure 3 – 1.

Fig. 3 – 1   Aircraft traction cascade system

The cascade system of aircraft and tractor was studied by the simulation method. In order to make the simulation results effective and consistent with the representation, it is necessary to design the parameters of the simulation system according to the actual situation, and the overall framework of the cascade system should be designed [101 – 103].

3. 1   Design of simulation function index

The simulation system consists of the following three parts: tractor simulation platform, aircraft simulation platform and traction cascade system motion platform.

The tractor simulation platform should have corresponding motion functions, and data transmission and processing functions. The motion functions include plane motion, traction and pushing, adjustable speed, adjustable turning radius, environment recognition, motion attitude adjustment, obstacle avoidance, adjustable light influence, on-line and off-line switching of the control mode and emergency stop function. The data transmission and processing functions include data video transmission, data in-depth analysis and processing, high-speed communication, network access, data interface expansion and multiple bus protocol access function. The other functions include IP68 level waterproof and dustproof (core control module), continuous motion control for multiple flights and quick charging, hardware expansion and software over the air (OTA) function.

The functions of the aircraft simulation platform include plane follow up function, center of gravity adjustment function, landing gear spacing adjustable function and weight adjustable function.

The motion platform functions of the cascade system include straight forward function, curve forward function, straight pushing function, curve pushing function, parking function, data transmission function and environment recognition function.

3. 2   Simulation parameters design

According to the *Technical Standards for Civil Airfields*, multiple scale simulation environments are constructed. The weight of the target aircraft is about 72500 kg, and the weight of its supporting tractor is about 16200 kg. According to the length simulation ratio of 1 : 30, the weight simulation ratio should be 1 : 2700, the weight of the aircraft simulation platform should be 2. 68 kg, the weight of the tractor simulation platform should be 0. 6 kg. The aircraft simulation platform used in the research has a weight of 2. 2 kg, and an additional iron piece counterweight is added to make its weight reach 2. 7 kg, and the traction simulation environmental platform has a weight of 0. 53 kg, both of which meet the simulation requirements.

According to the airport operation management regulations, the movement speed of the cascade system of the aircraft and the tractor should be adjusted according to different environments, and the maximum speed can not exceed 15 km/h. In the simulation experiments, the output power of the traction platform was adjusted to make the overall speed adjustable within the range of 0. 02 ~ 0. 15 m/s in the operation, which met the measurement requirements. The overall simulation environmental parameters is shown in Table 3 − 1.

Table 3 – 1 Simulation environmental parameters

Parameters	Range of parameters		
	The simulation ratio of 1 : 30	The simulation ratio of 1 : 100	The simulation ratio of 1 : 500
Weight of aircraft	2. 7 kg	—	—
Weight of tractor	0. 53 kg	—	—
Straight line speed of tractor	0. 02 ~0. 15 m/s	—	—
Tractor turning speed	0. 05 ~0. 2 m/s	—	—
Turning radius of tractor	0. 8 ~1. 5 m	—	—
Runway length	In line with a single full-process motion cycle	In line with a single full-process motion cycle	1 : 500 proportional zoom
Runway width	1. 36 m	In line with a single full-process motion cycle	1 : 500 proportional zoom
Runway slope	0° ~5°	0° ~5°	—
Runway friction coefficient	0. 26 ~0. 50	—	—
Aircraft turning angle	−60° ~ +60°	−60° ~ +60°	—
Tractor turning angle	−60° ~ +60°	−60° ~ +60°	—

3. 3 Simulation system architecture

The simulation environment was constructed with different simulation ratios, that is, at 1 : 30, 1 : 100 and 1 : 500 respectively, as shown in Figure 3 – 2. The simulation ratio should meet the design requirements of automatic control function indexes and parameter indexes.

(a)1 : 30   (b)1 : 100   (c)1 : 500

(a)1 : 30 for control function and communication function; (b) 1 : 100 for motion state;
(c)1 : 500 for data collection

Fig. 3 – 2   Multi-scale simulation platform

In the simulation study, the 1 : 30 simulation platform was used to analyze and verify the control function and communication function of the system; the 1 : 100 simulation platform was used to analyze and verify the traction cascade system motion process and the relationship of the motion state; the 1 : 500 simulation platform was used to collect data of the airport and various environments on the runway, such as aircraft, vehicles, signs, buildings.

In the field of aviation design, operational safety and reliability are emphasized and redundant design layers are added. The system framework consisting of operation layer, data transmission layer, control layer, model layer and data analysis layer was designed in the process of studying the multi-scale simulation platform.

3. 4   Research methods and technical path

3.4.1   Research methods

i . Comparative study method

The system function and performance with different scales, different models and different algorithms were compared, and the results were analyzed in order to select the most appropriate method for research.

ii . Analog control method

The analog control method was used in the control of the aircraft tractor, and the automatic control of the large passenger aircraft and the tractor was simulated by constructing multi-scale simulation, such as 1 : 30, 1 : 100, 1 : 500.

iii. Simulation experiment method

The simulation experiment method was used in the recognition of environmental information, and the content of image recognition was displayed in the form of images, and the environmental information was experimentally labeled.

iv. Quantitative method

The quantitative method was used in the high-speed data transmission in order to determine the transmission rate and the stability of high-speed data transmission.

3.4.2　Technical path

The technical path is shown in Figure 3 – 3. In terms of control, the machine learning network was constructed, the deep learning network was trained, the network parameters were adjusted, and the control of the aircraft tractor was output. In the image recognition, the image analysis processing was used to identify the information of aircraft, accompanying vehicles and staff, and it was used to evaluate the environmental information, the tractor path and the avoidance method.

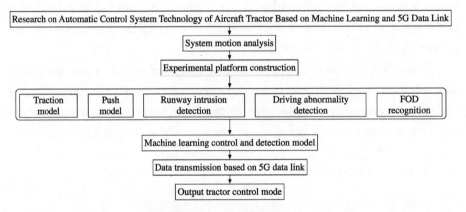

Fig. 3 – 3　Technical path

Chapter Four

PID Control and Stability of
Aircraft Traction Cascade System

4.1 Motion analysis and control mode of the cascade system

4.1.1 Motion analysis

The feasibility of applying machine learning to the tractor's automatic control can be verified by the stability and simulation study of the aircraft traction cascade system. The stability control method of the cascade system can be applied to the system in the form of upper-level knowledge. It can assist the decision-making process of the system and it plays a supporting role in the construction of machine learning networks and evaluation functions.

The motion process of the large passenger plane and the tractor is shown in Figure 4 – 1. The motion direction of the central axis of the passenger plane moves along with the motion direction of the central axis of the tractor in the whole motion process, and the motion direction of the central axis of the tractor is correlated to the direction of the steering wheel of the tractor.

Fig. 4 – 1　Aircraft traction process

The connection between the airplane and the tractor is shown in Figure 4 – 2. A is the airplane inclination angle(the angle between the central axis of the airplane and the horizontal axis) ; B is the tractor inclination angle(the angle between the central axis of the tractor and the horizontal axis) ; C is the front wheel steering angle(the angle between the direction wheel and the central axis of the tractor) ; (x,y) is the initial position of the airplane ; v is the motion speed.

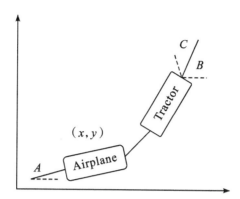

Fig. 4 – 2   Connection between the airplane and the tractor

The control objects of the system are the aircraft and the tractor, and the state variables are X, Y, A and B. The initial value is $X[0], Y[0], A[0]$ and $B[0]$ respectively, and the motion differential equations of the system can be written as Equation(4 – 1) to Equation(4 – 4).

$$\dot{X}_{(t)} = v\cos C(t)\cos B(t)\cos A(t) \qquad (4-1)$$

$$\dot{Y}_{(t)} = v\cos C(t)\cos B(t)\sin A(t) \qquad (4-2)$$

$$\dot{A}_{(t)} = k\cos C(t)\sin B(t) \qquad (4-3)$$

$$\dot{B}_{(t)} = -\dot{A}_{(t)} + k'\sin C(t) \qquad (4-4)$$

The aircraft and the tractor can be simplified as a cascade system according to Figure 4 – 2 and Equation(4 – 1) to Equation(4 – 4), and the steady state control method can be used to control the system.

4.1.2   Steady state control method

i . Proportional, integral, differential control

Proportional, integral, differential (PID) control methods can be divided into single loop PID control and dual-loop PID control. The single loop PID control can quickly realize the stability control of the cascade system as shown in Figure 4 – 2, but it can not correct the slight error between the steering wheel angle and the preset value during a single turn. As the tractor continues to move along the runway, the change value of the displacement and the angle will be accumulated, eventually causing the system to be unable to correct itself. The dual-loop PID controller executes the closed-loop control of the angle of the cascade system and the displacement of the tractor respectively. At this time, the deflection angle of the cascade system approaches zero degree, and the tractor displacement is a fixed value, thus achieving the stability control of the cascade system. In the steady state control of the cascade system, the state feedback control mechanism is introduced, and the integral and differential actions complement each other, which can reduce overshooting and increase the accuracy.

The key problem of PID control is parameter setting, which will directly affect the steady state process of the cascade system[104]. The parameter setting methods are empirical method and trial method. PID control relies on the model. Many nonlinear relations can not be linearized in the aircraft and tractor cascade system. Some parameters are ignored for linearization, which will cause a large error between the simulation and the actual system.

In the process of PID control, the parameters of maximum overshoot, rise time and static error are often used to evaluate the system. The maximum overshoot is used to measure the difference between the maximum peak value in the response curve and the steady-state value. It is an important indicator in the system stability evaluation process. The rise time refers to the time required for the response curve to reach the steady-state output value for the first time from the original working state. It is often used to evaluate the rapid response of the system. The difference between the given value of the controlled quantity and the stable value is called the static error, which is used to measure the accuracy of the cascade system.

ii . Linear quadratic regulator control

Linear quadratic regulator (LQR) control introduces optimal value in the classical control theory. The LQR control object of the aircraft traction cascade system is considered as the spatial state form of the linear system. The steady state control is obtained by solving the quadratic function of the cascade system.

In the control process, the state feedback controller is uniquely determined by the weight matrix. The optimal gain matrix can be solved by simulating the weight matrix, and the control method design can be completed. In the process of parameter adjustment, the parameter R is often reduced and the parameter Q is increased in order to reduce the adjustment time and the overshoot, and to increase the rise time for easy adjustment.

Due to the introduction of quadratic adjustment, the LQR control method can be well combined with genetic algorithm, particle swarm algorithm and other algorithms to obtain the optimal value of the parameter[105]. It can deal with nonlinear optimal control problems, and it can quickly obtain the optimal control of linear state feedback. LQR has the advantage of reducing the amount of calculation. Compared with the nonlinear control method, its calculation and implementation are relatively easy, and the control results can be applied to the nonlinear system under small signal conditions. In the fields of engineering application, the LQR control can achieve good dynamic performance and robustness, and it has gradually developed into a relatively mature control method in modern control theory.

iii. Sliding mode variable structure control

Sliding mode variable structure control is a mathematical method to reduce the dimensionality of the control process. In the control process, the sliding mode variable structure first generates a sliding mode surface in the state space by itself, then the controller controls the state variables and makes it move from the initial point to the sliding surface[106]. Driven by the sliding mode, the state variable continues to move on the sliding surface until the stability point stops. The most obvious difference of sliding mode variable structure control is the discontinuity compared with other control methods.

In the actual control process, the inertia of the system often causes the state variables to penetrate back and forth on the surface of the sliding membrane, which can not stop accurately at the stable position of the sliding surface. This type of penetration process is called chattering. High frequency chattering will seriously affect control accuracy and stability. Therefore, some advanced control technologies such as neural network control, adaptive control and fuzzy control are often used to assist the system to reduce chattering in the sliding mode variable control process[107 - 109].

iv. Fuzzy control

In the steady-state control process of the aircraft traction cascade system, new parameters such as distance, attendant personnel, vehicles, different airports and other environmental information are introduced constantly, and the data are constantly accumulating and iterating. As a result, the control of the entire cascade system will become more and more complex. In this case, the traditional classical control theory has its limitations, and the nonlinear and strong coupling of the complex system pose a new challenge to the precise system control.

Fuzzy control can deal with the complex cascade system well. Fuzzy control is based on the principle of incompatibility. When the complexity of the system increases, the clarity of the system description will decrease. When a certain threshold is reached, the complexity and clarity of the system will be mutually exclusive. The biggest difference between the fuzzy control and the classic control is that fuzzy control treats the cascade system as a black box, and the operation of the black box requires only a very small number of model parameters. According to the black box operation and its input and output, a certain control experience is formed and described as fuzzy rules. The steady state control of cascade system is realized by using fuzzy rules directly in simulation control. Fuzzy theory puts forward a set of effective methods, which can transform control experience, control rules and adjustment rules into mathematical functions, so that machines can read, recognize and use them.

4.1.3 PID control

The system adaptive steady state control was carried out for the cascade system composed of aircraft and tractor. The training samples and the upper level knowledge can be input into the machine learning network. The physical picture of the traction cascade system is shown in Figure 4 – 3, which shows the process in which the tractor pulls the airplane along the runway and pushes it into the hangar.

Fig. 4 – 3 Airplane towing vehicle

In the traction process, two actions are included, push and pull, as shown in Figure 4 – 4. During the pulling process, the motion of the aircraft will converge in the direction of the tractor motion. In the pushing process, the aircraft state is not positively correlated with the tractor motion with the change of the connection angle between the aircraft and the tractor. The aircraft will move to the other side when the tractor turns to one side over a certain angle. In order to realize the warehousing of the aircraft, the tractor should first turn in the opposite direction to form an initial deflection angle, and then constantly correct the angle deviation to form a new deflection angle in the opposite direction, so as to realize the aircraft steering process. Once the angle exceeds the threshold, the aircraft will move to the rear side and not to move forward.

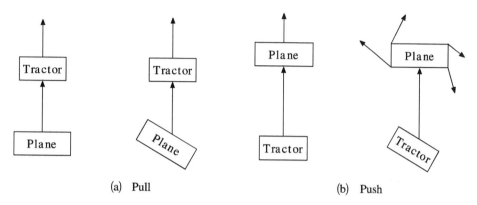

(a) Pull   (b) Push

Fig. 4 – 4   Traction actions

In these two processes, the PID is used to control and adjust the cascade system, so that the cascade system can move in the right direction and realize the overall stability control. The PID control process is shown in Figure 4 – 5.

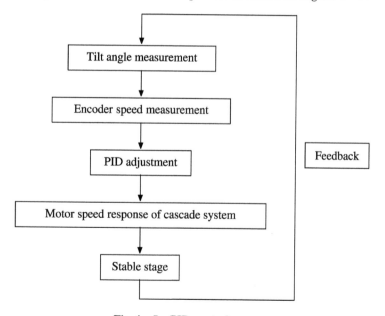

Fig. 4 – 5   PID control process

4. 2　Tilt angle measurement

For the traction cascade system, the set angle is input to the PID controller when the tractor pushes the aircraft into the hangar, and the cascade system goes straight or turns when receiving the output of the PID controller. At the same time, the differences between the steering angle and the central axis, steering angle and the angular acceleration feedback are compared with the set point, and continue to adjust, as shown in Figure 4 – 6.

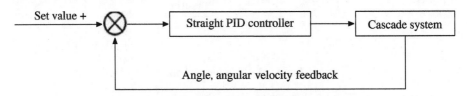

Fig. 4 – 6　Linear control of the cascade system

In the control process, it is necessary to use linear acceleration sensors and acceleration sensors to continuously measure the deflection angle, angular velocity, angular acceleration and motion trend of the point A in Figure 4 – 2. A linear acceleration sensor is used to measure the x, y, and z axial acceleration, as shown in Figure 4 – 7(a). The deflection angle of the aircraft traction cascade system can be calculated from the acceleration value on a single axis. The angular velocity can be calculated by the difference of the angle, and the actual output of the linear acceleration sensor is shown in Figure 4 – 7(b).

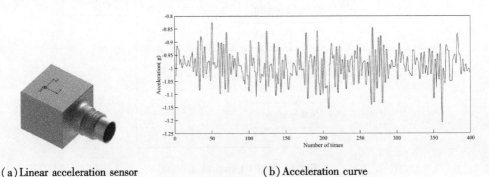

(a) Linear acceleration sensor　　　　　　(b) Acceleration curve

Fig. 4 – 7　Linear acceleration sensor and output curve

At the initial motion of the cascade system, the Y-axis acceleration can be calculated according to Equation $(4-5)$.

$$\text{Accel Y} = \sin \beta \cdot g \qquad (4-5)$$

Where,

β is the tilt angle of the aircraft;

g is the acceleration applied by the tractor.

The Y-axis acceleration of the tractor in actual motion is shown in Equation $(4-6)$.

$$\text{Accel Y} = \sin \beta \cdot g + \cos \beta \cdot \alpha \qquad (4-6)$$

α is the motion acceleration of the tractor, which is composed of the angular acceleration α_1 of the tractor and the linear acceleration α_2 of walking, as shown in Equation $(4-7)$.

$$\alpha = \alpha_1 + \alpha_2 \qquad (4-7)$$

During the operation of the cascade system, the difference between the motion acceleration and the pushing direction of the tractor is accumulated constantly. When only the linear acceleration sensor is used, the angle measurement error will increase due to the influence of the traction motion. In this case, angular velocity gyroscope should be used for processing. The gyroscope only outputs angular velocity and has low correlation with the motion direction of the cascade system, so it can deal with the system errors. The characteristic of the gyroscope and the accelerometer in the same motion process is shown in Figure $4-8$.

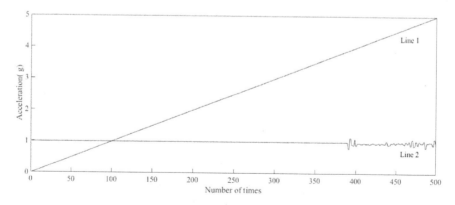

Fig. 4 –8   The feature of gyroscope and accelerometer

The top line (Line 1) in Figure $4-8$ represents the angle obtained by gyroscope integration, the bottom curve(Line 2) represents the axial angular velocity measured by the accelerometer. Equation $(4-8)$ is the complementary filtering equation.

$$angle\_A = K \cdot angle\_q + (1 - K1) \cdot (angle + gyro\_p \cdot dt) \qquad (4-8)$$

Where,

$angle\_A$ is the angle value after fusion;

$angle\_q$ is the angle of acceleration measurement;

$angle + gyro\_p \cdot dt$ is the angle obtained by gyroscope integration.

The common configuration parameters are as follows: the sampling period dt is 20ms; the weight value of the first-order complementary filtering is set to the weighted average; the wave filter coefficient $K1$ is set to 0. 02, which can realize the tilt angle measurement.

4. 3   Encoder velocity measurement

It is necessary to measure the motor speed before adjusting the cascade system. When the motion trend of the deflection angle is small, the motor needs to adjust its attitude slightly in order to make the system stable, and the rotational speed is low. However, when the motion trend of the deflection angle is large, the motor needs to adjust its posture considerably, with the rotational speed also high and the overshoot obvious. It is necessary to use photoelectric encoder to measure the speed of the DC motor accurately due to the small changes in the motion and the low recognition. The photoelectric encoder is composed of photoelectric detection device and code disc, as shown in Figure 4 - 9, it is a coaxial sensor. The motor drives the code disc to rotate, and the light source shoots from a fixed position to the code disc, and then passes through the detection grating to become a set of pulse signals. The motor speed can be calculated by counting the pulse signals per unit time, and two sets of square wave signals with a certain phase difference are used to judge the motor rotation. It is necessary for the simulation system to use the quadruple frequency technology to analyze the encoder signal in order to improve the control accuracy. The quadruple frequency method is used to measure the rising and falling edges of the A-phase and B-phase encoders at the same time. When the counting period is 4, the quadruple frequency technology can increase the sampling frequency by four times, making the number of pulse counts reach 16 times, and improving the accuracy of motor speed measurement, as shown in Figure 4 -9(c).

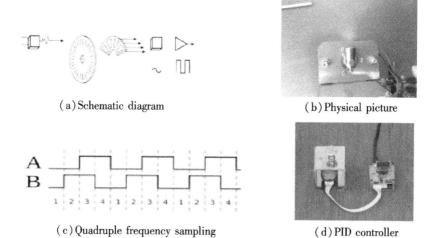

(a) Schematic diagram   (b) Physical picture

A
B

1 2 3 4 1 2 3 4 1 2 3 4

(c) Quadruple frequency sampling   (d) PID controller

Fig. 4 – 9   Photoelectric encoder

4. 4   DC motor PID control

PID adjustment is the process of proportional, integral and differential control of the system after identifying the deviation from the central axis and the current motor speed of the aircraft traction cascade system in the motion. The PID controller is shown in Figure 4 – 9 (d). The motion in different directions can be realized by the position closed-loop control and the speed closed-loop control of the DC motor, and the steady state maintenance of the cascade system can be achieved.

4.4.1   Position closed-loop control

Position closed-loop control is a control process in which the position information of the motor is derived according to the counting times of the photoelectric encoder and the control deviation is obtained by comparing with the expected value, and the deviation is reduced to zero gradually. It can be used to adjust the deflection angle in the traction cascade system. The discrete PID control form is shown in Equation(4 – 9).

$$\text{Pwm} = KP \cdot e(k) + KI \cdot \sum e(k) + KD[e(k) - e(k-1)] \qquad (4-9)$$

Where,

$e(k)$ is the current deviation;

$e(k-1)$ is the last deviation;

$\sum e(k)$ is the sum of the cumulative deviation;

Pwm is the output.

The position closed – loop control process is shown in Figure 4 – 10.

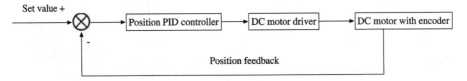

Fig. 4 – 10   Position closed-loop control

4.4.2   Speed closed-loop control

The pulse frequency per unit time is obtained based on the quadruple frequency sampling in speed closed-loop control. The motor speed is calculated and the control deviation is obtained by comparing with the target value. The deviation is controlled by PID and tends to zero gradually.

The speed closed-loop control can be used to control the direction and the speed of the entire system in cascade system, as shown in Equation(4 – 10).

$$\text{Pwm} = KP[\,F(k) - F(k-1)\,] + KI \cdot F(k) + KD[\,F(k) - 2F(k-1) + F(k-2)\,]$$
$$(4-10)$$

Where,

$F(k)$ is the current deviation;

$F(k-1)$ is the last deviation.

The speed closed-loop control process is shown in Figure 4 – 11.

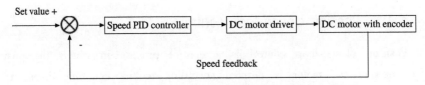

Fig. 4 – 11   Speed closed-loop control

4.5   Cascade system speed control and cascade PID

There are 27 motion states in the cascade system, that is, the current state of the tractor, aircraft, and cascade system in three directions of left, middle and right, and the desired motion state of the traction cascade system in three directions of left, middle and right, as shown in Figure 4 – 12. The key motion point of the cascade system is the angle of the system deviating from the axis equilibrium position. This angle is regarded as the deviation, and the speed control of the cascade system can be realized by using the negative feedback control to adjust the deviation until stable convergence.

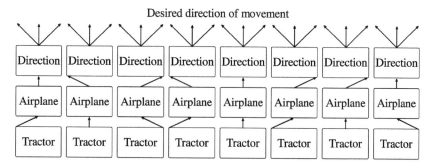

Fig. 4 – 12   Motion states in the cascade system

In the control process, the speed control is carried out before the axis stability control to ensure the priority of the cascade control, so that the axis stability of the cascade system will not be destroyed by the single speed negative feedback control, that is, the result of the speed control adjustment is only used as the target value of the axis change stability control. The principle of the cascade control system is shown in Figure 4 – 13.

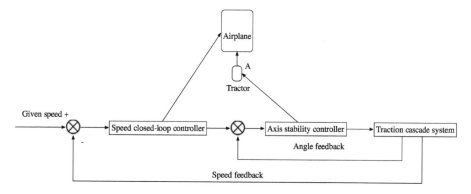

Fig. 4 – 13   The principle of the cascade control system

In Figure 4 – 13, the controller with stable angle and speed is connected in series, in which the output of speed control is used as the input of stability control, and the output of angle stability control is used as the output of the system, then the cascade control system is established. The composition of the system is shown in Figure 4 – 14.

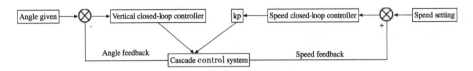

Fig. 4 – 14   Composition of the cascade control system

The angle stability control algorithm and the speed control algorithm are shown in Equation(4 – 11) and Equation(4 – 12) respectively.

$$a = KP \cdot (\theta - \alpha_1) + KD \cdot \theta' \qquad (4-11)$$

$$a_1 = KP1 \cdot e(k) + KI1 \cdot \sum e(k) \qquad (4-12)$$

Where,

θ is the angle;

θ' is the angular velocity;

$e(k)$ is the speed control deviation;

$\sum e(k)$ is the integral of the speed control deviation.

In order to further simplify the control system, Equation(4 – 11) and Equation (4 – 12) are combined to obtain Equation(4 – 13).

$$a = KP \cdot \theta + KD \cdot \theta - KP[KP1 \cdot e(k) + KI1 \cdot \sum e(k)] \qquad (4-13)$$

The speed control of the cascade system can be realized theoretically by the iterative solution of Equation(4 – 13).

4. 6   PID traction control simulation

4.6.1   Simulation experimental method

In the simulation experiment, the angular acceleration sensor was used to identify the attitude of the cascade system, and the attitude information was input into the PID controller to control the operation of the motor. The axis deflection angle of the cascade system was approaching zero by adjusting the working mode of the motor. The modified angle information was fed back to the PID control, and the steady state was maintained by constant adjustment.

The method of repeated test and simulation was used in the experiment. According to the control process, as shown in Figure 4 – 15, the steady-state control at different initial angles and the steady-state maintenance at different velocities were studied.

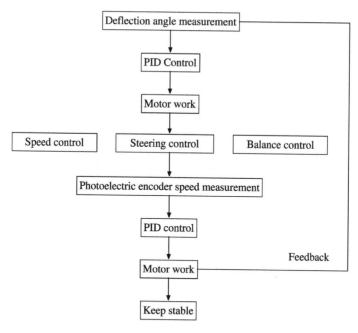

Fig. 4 – 15   Cascade system stability control process

In order to improve the control accuracy and increase the speed information per unit time, the sampling period was reduced to 5 ms, the weight of the first-order complementary filtering was still set to the weighted average, and the wave filter coefficient $K1$ was adjusted to 0. 05.

4.6.2   Simulation experiments

i . Building the runway simulation environment

The gray scale was RGB(128,128,128), the simulation runway spacing was 1. 2 m, and the width of runway line was 5 cm, as shown in Figure 4 – 16.

Fig. 4 – 16   The simulation runway

ii . Construction of cascade system verification platform

The verification platform of the cascade system is composed of cascade system simulation module, tilt angle measurement module, motor drive module, PID control module and auxiliary steering module, as shown in Figure 4 – 17.

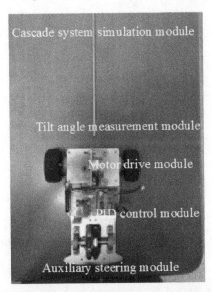

Fig. 4 – 17   Verification platform of the cascade system

iii. Cascade system stable maintenance at different initial angles

The initial angle of the cascade system was adjusted, and the steady-state control process was tested when the initial angle of the cascade system was $-45°$, $-70°, 0°, 75°$ and $45°$ respectively. The starting state at different initial angles is shown in Figure 4 – 18.

Fig. 4 – 18   Starting state at different initial angles

The real-time deflection angle was input to the PID control module at different initial angles in the PID control process. The speed of motor was adjusted to control the steering system, and the deflection angle was reduced continuously. The motion state in the control process is shown in Figure 4 – 19, where the deflection angle is approaching zero.

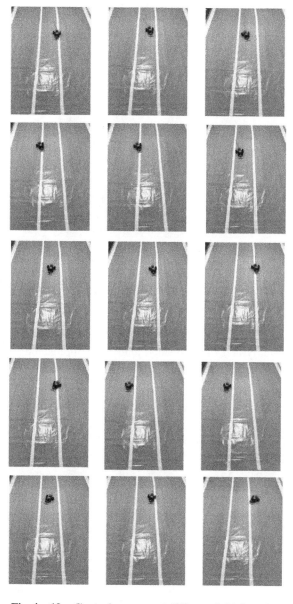

Fig. 4 – 19   Control process at different initial angles

iv. Cascade system stable maintenance at different initial speeds

The speed was controlled by the square wave signal. The initial speed ratio was set as 10% ,20% ,30% ,50% ,100% respectively, and the starting state at different initial speeds is shown in Figure 4 – 20.

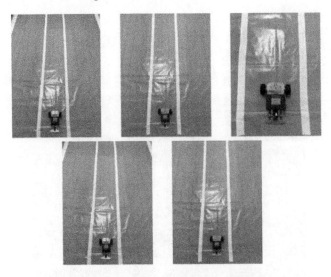

Fig. 4 – 20   Starting state at different initial speeds

The real-time deflection angle was input to the PID control module at different initial speeds in the PID control process. The motor speed was adjusted, and the deflection angle was reduced gradually. It is a control process in which the speed was adjusted constantly and the deflection angle was approaching zero, as shown in Figure 4 – 21.

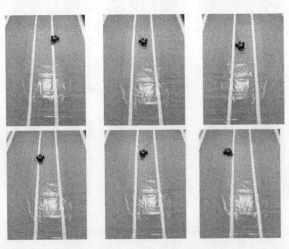

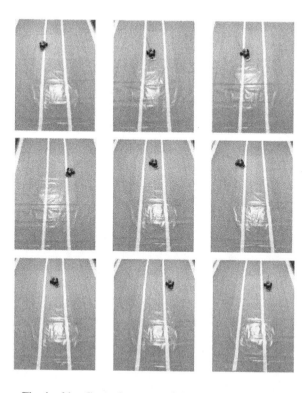

Fig. 4 – 21   Control process of different initial speeds

ⅴ. Parameter tuning

The PID parameters were adjusted to make the cascade system move in the direction of the optimal feedback value. In the experiment, the PID parameter tuning process was as follows. Firstly, the value of I and D was set to zero. Secondly, the value of P was increased gradually from zero to the system oscillation, and $KP = 500, KI = 0, KD = 0$. Finally, PID parameters were set as $KP = 50, KI = 0, KD = 0$ and $KP = 500, KI = 0, KD = 400$, respectively.

4.6.3   Simulation experimental results

ⅰ. Experiment at different initial angles

The experiment was repeated 200 times, and some experimental results at different initial angles are shown in Figure 4 – 22.

The repeated test results at the initial angles of $-45°$, $-70°, 0°, 75°$ and $45°$ are shown in Table 4 – 1. The experiment was repeated 20 times for each group. The control results at different initial angles are shown in Figure 4 – 23.

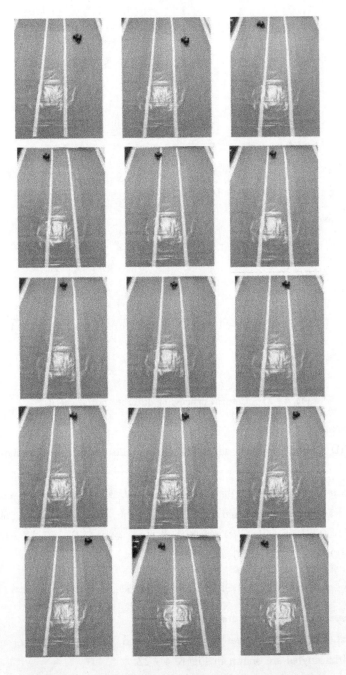

Fig. 4 – 22   Experimental results at different initial angles

Table 4 –1   Repeated test results at different initial angle

Group number	Initial angle / °	Average running time / s	Average deflection angle / °	Average vertical deflection angle / °
1	-45	9.0	4.8	44.2
2	-45	8.2	11.3	47.2
3	-70	9.2	1.2	12.7
4	-70	8.4	3.0	89.6
5	0	10.9	13.9	60.9
6	0	10.2	4.7	33.9
7	75	10.9	10.3	6.1
8	75	8.9	9.3	38.1
9	45	11.0	19.1	11.2
10	45	8.5	9.4	10.6

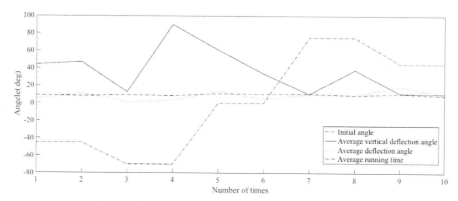

Fig. 4 – 23   Control results at different initial angles

Under the action of PID feedback regulation in the motion control process, the output waveform of motor and the waveform of deflection angle at different initial angles are shown in Figure 4 – 24.

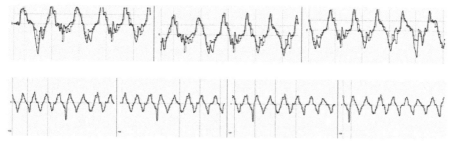

Fig. 4 – 24   Waveforms of motor output and deflection angle at different initial angles

In the control process of different initial angles, the average deflection angle was obviously approaching zero, and the motion state of the system had a high fitting degree with the angular acceleration sensor and the motor output waveform, which indicated that the PID control process was effective and it could control the cascade system. Since the PID regulation was only related to the deflection angle of the cascade system, the cascade system did not move along the middle line of the runway in the experiment.

ii. Different initial speeds

The experimental results at different initial speeds are shown in Figure 4 − 25.

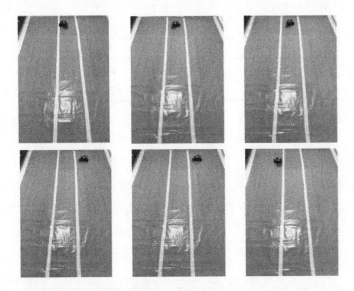

Fig. 4 − 25　Experimental results at different initial speeds

The initial speed ratio was 10% ,20% ,30% ,50% and 100% respectively in the experiment, with the results shown in Table 4 − 2. Each group of experiments was repeated 20 times. The control results at different initial speeds are shown in Figure 4 − 26.

Table 4 − 2　Control results at different initial speed ratios

Group number	Initial speed ratio / %	Average running time / s	Average deflection angle / °	Average vertical deflection angle / °
1	10	11. 3	8. 8	69. 1
2	10	9. 7	1. 3	11. 4

(Continued)

Group number	Initial speed ratio / %	Average running time / s	Average deflection angle / °	Average vertical deflection angle / °
3	20	10. 1	2. 0	78. 9
4	20	9. 9	4. 3	55. 7
5	30	7. 9	11. 4	46. 6
6	30	8. 7	4. 9	26. 5
7	50	7. 7	4. 0	44. 9
8	50	7. 9	5. 8	23. 4
9	100	2. 3	82. 2	70. 8
10	100	3. 1	87. 1	86. 9

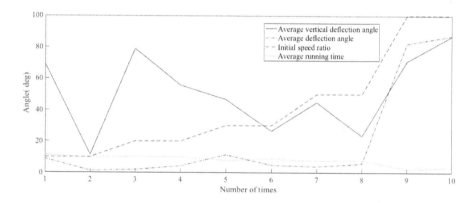

Fig. 4 – 26   Control results at different initial speeds

Under the action of PID feedback adjustment, the waveforms of motor output and deflection angle at different initial speeds are shown in Figure 4 – 27.

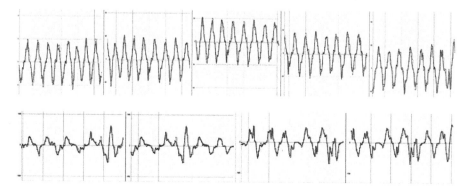

Fig. 4 – 27   Waveforms of motor output and deflection angle at different initial speeds

In the control process at different initial speeds, the average deflection angle is related to the speed, and the deflection angle of the system can approach zero when it is lower than the threshold value. The motion state of the system has a high fitting degree with the angular acceleration value and the motor output waveform, which indicates that the PID control process is effective and it can control the cascade system. However, when the speed ratio exceeded 50%, since the PID adjustment could not keep up, the system broke and was locked directly. Since the PID regulation was only related to the deflection angle of the cascade system, the cascade system did not move along the middle line of the runway in the experiment.

iii. Parameter tuning

The value of I and D was set to zero, and the value of P was being increased gradually from zero to the system oscillation, $KP = 500$, $KI = 0$, $KD = 0$. The response curve is shown in Figure 4 – 28.

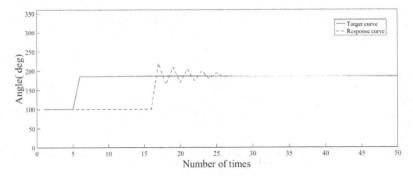

Fig. 4 – 28　Response curve($KP = 500$, $KI = 0$, $KD = 0$)

When $KP = 50$, $KI = 0$ and $KD = 0$, the response curve is shown in Figure 4 – 29.

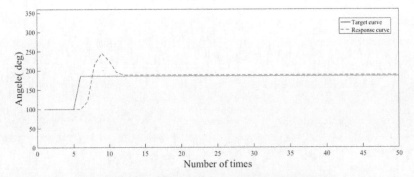

Fig. 4 – 29　Response curve($KP = 50$, $KI = 0$, $KD = 0$)

When $KP = 500$, $KI = 0$ and $KD = 400$, the response curve is shown in Figure 4 – 30.

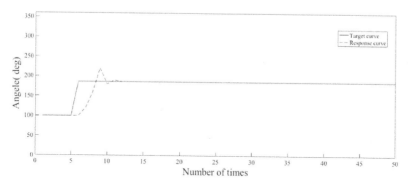

Fig. 4 – 30   Response curve ($KP = 500$, $KI = 0$, $KD = 400$)

It can be seen that the PID adjustment can be used to realize the stability control of the cascade system. The value of P is positively correlated with the oscillation. The cascade system has a static error when the value of P is small, and the response speed is obviously reduced. In some cases, the static error can be eliminated by increasing the value of P, and the response speed of the system can be improved. However, an excessively high value of P will cause the system to enter oscillation. It is necessary to increase the differential control in order to suppress the oscillation effectively. The differential control increases the system damping and will reduce the oscillation impact and the response speed of the system. Therefore, it is necessary to optimize the system by combining the value of P and I.

In the position control, the control of P was used at the beginning, and the coefficient of P was increased until the system was oscillation, then the differential control was added to increase the damping. After eliminating the oscillation, the parameters of P and I were adjusted according to the specific requirements of the system for response and static difference. Generally, the control difficulty of a control system depends on the moment of inertia of the system and the requirement for response speed. The requirement for the response speed is lower when the moment of inertia is small, and the PID parameters will be less sensitive.

4.7   Conclusions

The stability simulation of the aircraft traction cascade system was studied. The stability control of the aircraft traction cascade system was realized by PID control, linear quadratic regulator control, sliding membrane variable structure control and fuzzy control.

The stability of the aircraft traction cascade system was studied by tilt angle measurement, encoder speed measurement, motor PID control, speed control and cascade PID control. The stability control of the cascade system could be realized in the condition of small inertia.

Motor speed and angle waveform in stable holding, straight running and initial angle control were studied by the stability control experiment. Through the numerical fitting of the angular acceleration sensor, the motion state of the cascade system was stable, and the tractor could be driven steadily on a straight line. When it was being driven on a curve, the speed could be reduced and the stable control was achieved.

The response characteristics of the system were studied by PID parameter adjustment and the stability control of the cascade system was realized. The value of P was positively correlated with the oscillation. The response speed could be improved by increasing the value of P, but it caused the system oscillation. The oscillation of the system could be restrained by the differential control, but the response speed of the system was obviously slow.

Chapter Five

Motion Control of Aircraft Traction
Cascade System Based on Machine Learning

At present, the driver always operates the tractor manually to complete the aircraft towing and pushing process in a relatively stable environment at airport. Although the control process and environment information are numerous and complex, they are relatively easy to be recognized with machine learning. The machine learning network can be established with image recognition[110], environmental feature detection, motion control information, etc. The process of the tractor pushing the aircraft into hangar and towing the aircraft can be abstracted into the following machine learning items:

(1) Collect data such as runway line image, turning angle, motor speed, and environmental characteristics.

(2) Obtain the tractor motion control experience with manual operation.

(3) Use classification methods to divide the data into training set and test set.

(4) Perform feature extraction and selection on the training set, and determine the task to be processed.

(5) Build a machine learning network.

(6) Measure the experience in the process of completing the task.

(7) Adjust the machine learning network to improve the operational measurement.

The method of machine learning was used to study the simulation motion of towing and pushing after the establishment of the simulation environment. There are many steps in the process of using machine learning to realize the automatic control of the tractor and aircraft cascade system, as shown in Figure 5 – 1. Firstly, the motion process of the cascade system was abstracted into a mathematical problem, and a supervised learning model was constructed by the machine learning network. Secondly, the model data processing and parameter tuning were carried out to realize the simulation control of the cascade system.

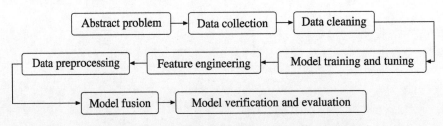

Fig. 5 – 1　Machine learning steps of cascading system

The internal links of the cascade system are complex, and the motion control process of the cascade system can be classified into two types, i. e. motion control and environmental restraint control. Motion control can be divided into traction control and pushing control. In the traction process, the steering and power of the motor and the tilt angle of the cascade system were approximately considered to have a linear relationship, and the convolutional neural network was used to analyze the image and control the steering. The motor steering and power had a non-linear relationship with the tilt angle of the cascade system in the pushing process which was affected by many related parameters. A neural network with multiple types of inputs was introduced to recognize the feature of inputs, such as image, angle, acceleration, deflection, speed, power, runway deviation, and the corresponding machine learning model was obtained. While the tractor is pushing the aircraft into the hangar, the influence of offset angle, traction direction, tire steering and other parameters on the system motion is nonlinear, and the state of the system will not be able to correct itself after the threshold is exceeded. The completely different motion states of the cascade system will pose new challenges to the existing control methods. Therefore, a method based on machine learning was proposed, and a large number of data samples were used to establish a machine learning model to complete the automatic control of the cascade system under nonlinear and multi-state conditions. Environmental restraint control was used to identify the motion environment of the cascade system, and to avoid or stop the motion when runway intrusion, abnormal driving state, FOD and other special conditions were detected.

5. 1   Motion control based on machine learning

5.1.1   Motion analysis

The aircraft motion was controlled by operating the tractor manually, and the attitude information of the aircraft and the tractor during the motion was obtained. A 1 : 100 simulation model was used for the tractor and aircraft, and attitude sensors were installed, as shown in Figure 5 – 2.

Attitude sensor configuration requirements:

Voltage:3. 3 V ~5 V;

Current: <40 mA;

Size:51. 3 mm ×36 mm ×15 mm.

Measurement dimension: acceleration, angular velocity and angle are all three

Fig. 5 – 2   Motion attitude recognition

dimensions.

Range: acceleration ± 16 g, angular velocity $\pm 2000°/\text{s}$, angle x,z $\pm 180°$, y axis $\pm 90°$.

Angle accuracy: dynamic $0.1°$, static $0.05°$.

Sampling rate: 100 Hz.

Baud rate: 115200 bps.

i . Traction analysis

Traction attitude recognition process is as follows: The traction motion model was analyzed when the relation of angle and position of the tractor, then aircraft and runway was changed. Both the tractor and the aircraft were equipped with acceleration sensors, and the tractor was controlled manually to perform S-shaped movement during the traction process, as shown in Figure 5 – 3.

Fig. 5 – 3   Traction motion at different times

The comparisons of acceleration, angular velocity and angle recorded by the acceleration sensor during the traction simulation process are shown in Figure 5 – 4.

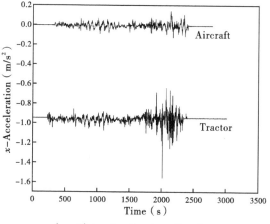

(a-1) Acceleration in x direction

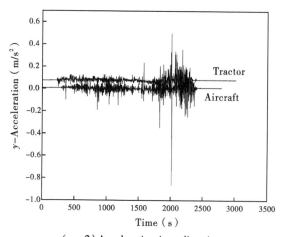

(a-2) Acceleration in y direction

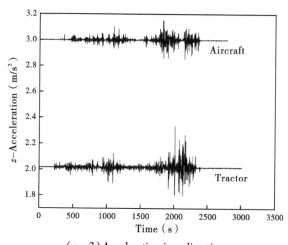

(a-3) Acceleration in z direction

(a) Acceleration

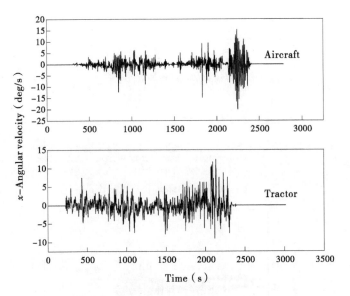

(b – 1) Angular velocity in *x* direction

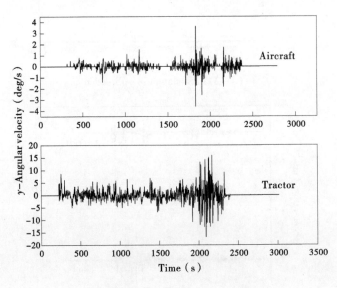

(b – 2) Angular velocity in *y* direction

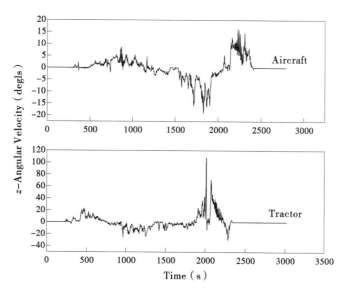

(b-3) Angular velocity in z direction

(b) Angular velocity

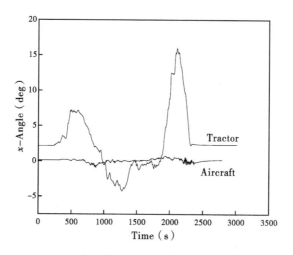

(c-1) Angle in x direction

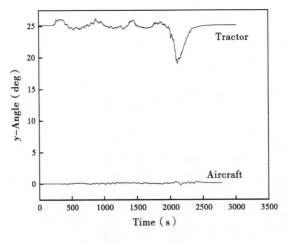

(c – 2) Angle in y direction

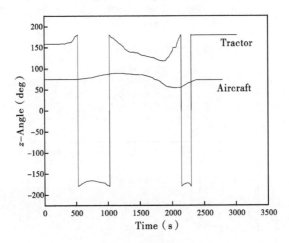

(c – 3) Angle in z direction

(c) Angle

Fig. 5 – 4   Comparisons of traction simulation process

The motion attitude of the tractor has a high degree fit with the attitude of the aircraft, as seen in the motion posture of X and Y axis. It can be simplified to a linear regression model, i. e. , the aircraft will follow the tractor when it moves. When approaching the runway centerline, the tractor needs to cross the runway line and vibrate slightly along the runway centerline in order to keep the aircraft in a relatively stable motion state. The convolutional neural network was used for model construction in machine learning.

ii . Pushing process analysis

The tractor is behind the moving direction of the cascade system during pushing process, but the tractor is in front of the cascade system during the traction process, so the motion control of these two modes is completely different. It is necessary to identify whether the cascade system is in a pushing state in the process. Compared with the traction process, the biggest difference of the images collected in the pushing process is that the characteristics of aircraft tires can be identified. Whether the aircraft wheels can be detected is used as a basis for judging the current tractor in the pulling or pushing mode. The pushing process at different moments is shown in Figure 5 – 5.

Fig. 5 – 5　The motion state of pushing at different times

The comparisons of acceleration, angular velocity and angle recorded by the acceleration sensor during the pushing simulation process are shown in Figure 5 – 6.

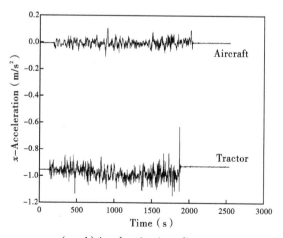

(a – 1) Acceleration in x direction

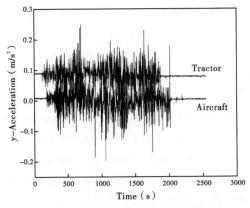

(a − 2) Acceleration in y direction

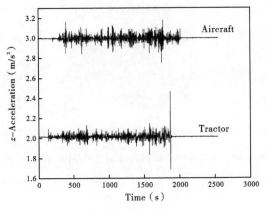

(a − 3) Acceleration in z direction

(a) Acceleration

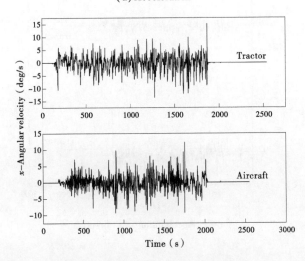

(b − 1) Angular velocity in x direction

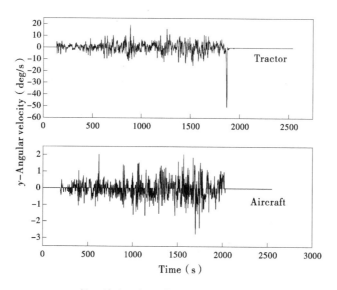

(b − 2) Angular velocity in y direction

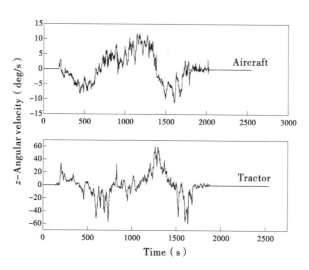

(b − 3) Angular velocity in z direction

(b) Angular velocity

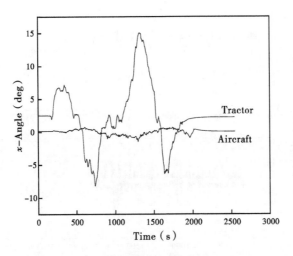

(c - 1) Angle in x direction

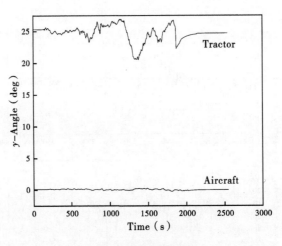

(c - 2) Angle in y direction

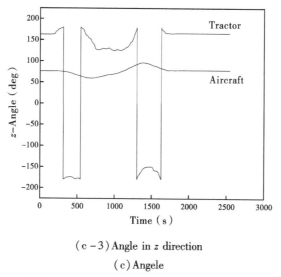

(c - 3) Angle in z direction

(c) Angele

Fig. 5 – 6   Comparisons of pushing process

Figure 5 – 6 shows that the motion relationship between the aircraft and the tractor is nonlinear in the pushing process, and its features include the characteristics of the aircraft wheel, the runway image and the attitude characteristics of the aircraft and the tractor. The model was constructed by using multiple types of input neural network in supervised learning.

5.1.2   Automatic control model

After analyzing the cascade system, two kinds of models of pulling and pushing were set up for automatic control according to the motion rule and the state of the cascade system.

i . Data collection

The model needs to extract the features, such as environmental information and the system's posture information, while the tractor is pushing the aircraft into the hangar. The environmental information includes video images taken by the front camera of the tractor, ground runway images, aircraft views, straights, curves, steering rates, steering angles, etc. The attitude information obtained by the acceleration sensor includes aircraft attitude, motor operating conditions, acceleration sensor values, cascade system speed, tractor attitude, cascade attitude, etc. , as shown in Figure 5 – 7.

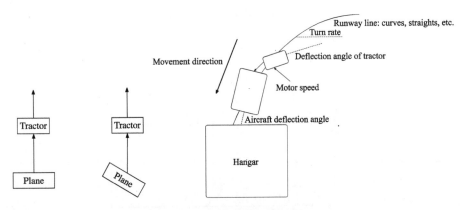

Fig. 5 – 7   Data acquisition process

The characteristic information of the cascade system will be changing in real time with the motion of the cascade system. Machine learning was used in the process, and the training analysis was carried out to generate an automatic control model for controlling the movement of the cascade system after the data samples were obtained by manual operation. In the control process, the extraction of all features will cause many problems, such as too many dimensions, long training time and system overfitting, due to the complex features and multiple dimensions. Feature selection is a process of selecting a subset related to the current learning task from all features. For example, the video image background has a large range of the same area in some special scenes, such as outfield bases, and sparse learning can be used to select features to reduce the complexity and dimension of features. Therefore, the selection of feature subsets will directly affect the training results of the learner. In the research, feature selection was carried out in the way that the feature was easy to identify, that the feature had physical meaning, and that the feature change was locally controllable.

ii. Traction model construction

The traction process consists of straight traction and curve traction. The aircraft's motion direction has a delayed positive correlation with the tractor motion direction, that is, when the tractor is moving in a predetermined direction, the aircraft will first move in the opposite direction for a certain distance and then move in the predetermined direction if the inclination angle exceeds the threshold. The whole system maintained convergence in this process, and the process was analyzed and the traction model was designed, which was composed of convolutional neural network.

A convolutional neural network is composed of an input layer, two pooling layers, two convolutional layers, two full connection layers and one output layer. The runway images collected during traction were put into the convolutional neural network, and the angles were annotated for different runway extraction results. The annotation content was the corresponding output, and the output of the convolutional neural network was the offset angle of the runway, which was set according to the different steering angle during operation. The existing data samples were trained to generate the results. The identification process of straight lanes is shown in Figure 5 – 8, and the identification process of curves is shown in Figure 5 – 9.

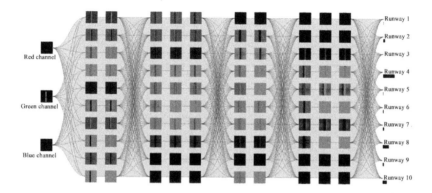

Fig. 5 – 8   Linear running recognition process

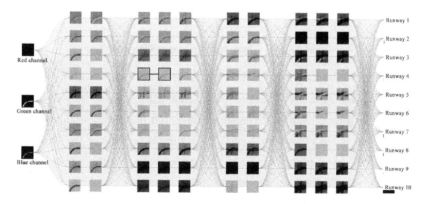

Fig. 5 – 9   Curve driving identification process

The calculation of convolutional neural network is shown in Equation(5 – 1).

$$s(i,j) = \sum_m \sum_n x(i - m, j - n) w(m,n) \qquad (5 - 1)$$

Where,

x is the input;

w is the convolution kernel;

$i = 1, 2, \cdots, m$, and $j = 1, 2, \cdots, n$.

iii. Pushing model construction

The environmental information of pushing model includes wheel detection information, runway image information, tractor attitude information, aircraft attitude information, etc. Data categories include category value data, image data, posture data, etc. The pushing model has many input types, and the output is the steering angle of the tractor. The neural network model with multi-category input was designed, and its architecture is shown in Figure 5 – 10. The model can receive various types of input information such as images, category value and feature data value, and the fitting output was obtained through machine learning.

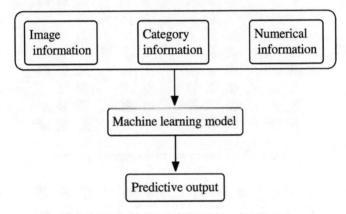

Fig. 5 – 10   Multi-category input framework

The neural network model of the pushing process is shown in Figure 5 – 11. In this model, image data is 160 dimensional data, the attitude information of the tractor and the aircraft is nine dimensional data, and the attitude information of the cascade system is ten dimensional data. The automatic control of the pushing model includes the following processes. Firstly, the network operation parameters are fitted manually to generate the pushing control model. During the pushing process of the cascade system, the traction module recognizes whether there is aircraft wheel information in the image and judges whether the pushing model is in the motion control state. Secondly, the collected image information, motion state category information and attitude information of the cascade system are transmitted to the neural network with multi-category input. The network performs preprocessing, standardization, dimensionality reduction, concatenation and full connection of multi-category data, where all of the connection parameters are consistent with the training model. Thirdly, output is transmitted to the steering gear to realize the pushing motion of the cascade system.

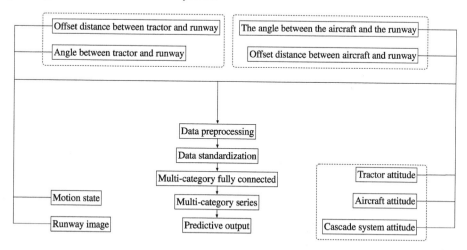

Fig. 5 – 11   Multi-category input neural network

iv. Model calculation

After the construction of the motion model and the feature selection are completed, the gradient and the back propagation methods are used to analyze and calculate the model.

The gradient descent method is an algorithm commonly used in machine learning to find the global minimum point of the objective function. The algorithm searches for the optimal solution along the direction where the function value drops fastest, that is, the negative gradient direction. When the algorithm iterates to a certain point where the gradient is zero, the parameter update will stop, and it is considered that a local minimum point has been searched. The commonly used gradient descent methods include BGD batch gradient descent method, SGD stochastic gradient descent method and momentum method.

In BGD scale gradient descent method, the loss function and gradient calculation of the entire training set are very slow, and there are fewer parameters to set, so BGD is not suitable for the analysis and processing of big data samples. The BGD algorithm uses the data of the full training set for gradient update calculation, as shown in Equation(5 – 2).

$$\theta = \theta - \eta \cdot \nabla_\theta J(\theta) \qquad (5-2)$$

When the SGD stochastic gradient descent method updates and calculates the parameter gradient, it simply uses a single or a small number of training samples to estimate the expected value, as shown in Equation(5 –3).

$$\theta = \theta - \alpha \cdot \nabla_\theta J(\theta; x^{(i)}, y^{(i)}) \qquad (5-3)$$

Where,

$(x^{(i)}, y^{(i)})$ is a sample in the training set.

The momentum method is suitable for dealing with some problems which solution is a minimum value rather than an optimal value. The momentum method is a way to push the value of the objective function toward the optimal solution quickly, the update process of momentum is shown in Equation(5 –4).

$$v = \gamma v + \alpha \cdot \nabla_\theta J(\theta; x^{(i)}, y^{(i)}), \theta = \theta - v \qquad (5-4)$$

Where,

v is the current velocity vector(parameter update amount), the same dimension as the parameter vector θ.

α is learning rate, and the value of the momentum must be small because of the large gradient magnitude while using momentum.

Backpropagation passes the error to the hidden layer of the neural network. In the neural network with multiple inputs, the back propagation method is used to calculate the parameters of each level. The connection weight of the backpropagation output layer and the hidden layer is adjusted according to the error, and the threshold is determined. The mean square error of the hidden layer neurons is calculated, and the connection weight value and the threshold value of the input layer and the hidden layer are adjusted according to the error, as shown in Figure 5 –12.

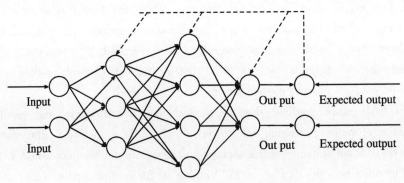

Fig. 5 –12　Back propagation calculation

The algorithm carries on the forward transfer process, then the input signal from the input layer of the BP network passes through the hidden layer to calculate the output value of each neural unit in turn. The output error in the forward transfer process is fed back layer by layer through the back propagation process, and the weight value of the input layer and hidden layer is corrected according to the error input in the second step. The calculation steps are as follows:

(1) Calculate the output of each level unit in the forward process according to Equation(5 –5).

$$Out_m : Out_m = (1 + e^{-\sum_m w_{mn} o_m})^{-1} \tag{5-5}$$

(2) Calculate the gradient information of the output layer according to Equation(5 –6).

$$\partial_m : \partial_m = (1 + e^{x_i})^{-1} Out_m (1 - Out_m) \tag{5-6}$$

(3) Calculate the gradient information of each hidden layer in reverse according to Equation(5 –7).

$$\partial'_m : \partial'_m = Out_m (1 - Out_m) \sum_n w_{mm} \partial_n \tag{5-7}$$

(4) Calculate the weight correction amount after back propagation according to Equation(5 –8).

$$\Delta w_{mn}(t) = a\Delta w_m n(t-1) + b\partial_m Out_n \tag{5-8}$$

(5) Revise the weight value of the original transmission process according to Equation(5 –9).

$$w_{mn}(t+1) = w_{mn}(t) + \Delta w_{mn}(t) \tag{5-9}$$

The backpropagation network solves the problem of hidden layer weight correction. It is easy to fall into the local minimum point and cannot converge to the global minimum point when using the gradient as the error correction parameter. The common loss functions are shown in Table 5 – 1.

Table 5 –1   Loss functions

Category	Characteristic	Function
Logistic regression	Logistic regression and softmax classification are performed using mutual entropy loss and logarithmic loss.	$MSLE : L = \dfrac{1}{n} \sum\limits_{i=1}^{n} (\log(y_{true}^{(i)} + 1) - \log(y_{pred}^{(i)} + 1))^2$
Square loss	The least squares formula is derived by maximum likelihood estimation and the loss value is calculated.	$MSE : L = \dfrac{1}{n}(y_{true}^{(i)} - y_{pred}^{(i)})^2$

(Continued)

Category	Characteristic	Function
Hinge loss function	Hinge loss variance and hinge loss are used for calculation, which is mainly used in support of vector machines.	$L(m_i) = \max(0, 1 - m_i(w))$ $L = \frac{1}{n}\sum_{i=1}^{n}(\max(0, 1 - y_{true}^{(i)} \cdot y_{pred}^{(i)}))$
Other loss functions	Include $0 \sim 1$ loss, absolute loss, Poisson loss and cosine similarity evaluation.	$\text{MSLE}: L = \frac{1}{n}\sum_{i=1}^{n} \mid (y_{true}^{(i)} - y_{pred}^{(i)}) \mid$ $L = \frac{1}{n}\sum_{i=1}^{n}(y_{pred}^{(i)} - y_{true}^{(i)} log(y_{pred}^{(i)}))$ $L = \dfrac{\sum_{i=1}^{n} y_{pred}^{(i)} \cdot y_{true}^{(i)}}{\sqrt{\sum_{i=1}^{n}(y_{true}^{(i)})^2} \cdot \sqrt{\sum_{i=1}^{n}(y_{true}^{(i)})^2}}$

5.2 Environment constrained motion control based on machine learning

Planes take off and land frequently in large civil aviation airports, and the field service vehicles are busy. The risk of safety accidents increases during aircraft towing or propulsion operation in the complex environment composed of aircraft vehicles and personnel, which will affect passengers and airport safety. The test plane often needs to transfer at various civil airports in the flight test. Various vehicles, such as food delivery truck, snow removal truck, sewage truck, air-conditioned truck, fuel truck, shuttle truck, and shuttle in the airport, are moving forward and backward, departing or approaching the aircraft, so the risk of collisions or other accidents between cars and people is high. In addition, foreign objects on the runway can also pose a safety hazard to the airport. After using machine learning to realize tractor simulation automatic control, the tractor needs to recognize the relevant runway information of the environment to assist the control process.

During the motion of aircraft and tractors, especially some prototype models, accompanying personnel, such as field and quality control personnel and other operators, and accompanying vehicles, such as towing, refueling, lifting, and transportation vehicles are complicated. At the same time, different types of ladder cars are used frequently and the hangar environment is complicated, as shown in Figure 5 – 13. The traction cascade system needs to recognize the environment for path planning and obstacle avoidance in the aircraft traction process.

Fig. 5 – 13   Actual motion scene

The motion of the system was reported by the camera mounted on the tractor, and the runway intrusions were tracked according to the target shape, speed, distance, quantity and distribution of the surrounding environment. The motion state of the aircraft tractor was monitored effectively by image processing. When the runway intrusion matches the pre-set dangerous speed and distance, the tractor will be alerted and displayed on the display.

During the motion, the environmental information consists of runway, vehicles, accompanying personnel, tower, buildings, other aircraft, ground signs, runway lights and so on. In the process of automatic control, feature detection, water mist model detection, block detection and other relevant algorithms are used to quickly and intelligently recognize images and identify environmental information and conditions, so as to build an overall environmental auxiliary judgment system, as shown in Figure 5 – 14.

The road surface condition of abnormal driving and the state of motion are detected and identified by the feature detection in the motion. The block image processing method is used to quickly identify the runway intrusion in the airport. The FOD on the runway is detected by the water fog model detection method. The results are fed back to the tractor control by wireless transmission for timely avoidance and avehicle stopping.

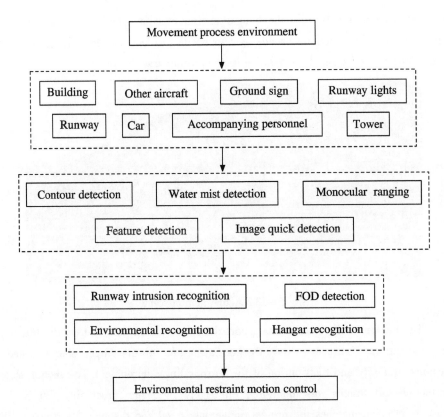

Fig. 5 – 14   Motion environment detection

5.2.1 Characteristic environmental constraint

The tractor is located in the runway centerline in the normal motion of the cascade system. The image information is composed of runway, guide lines and related signs, and its feature is single. When the tractor deviates from the center line of the runway or travels to an abnormal area, the environmental information contains a lot of other characteristic information. So the tractor can be stopped in an abnormal motion state through the identification of the number of feature points. The feature detection methods in the motion of the cascade system are as follows.

i . Edge detection

Edge detection is a commonly used multi-level edge detection algorithm, which can effectively identify the actual edge of the object in the image and reduce the probability of missed detection and false detection of the image edge[111]. Edge detection can identify the actual edge point closest to the edge position, and the fitting effect is excellent[112-113]. The detection process mainly relies on the four parameters, i. e. position, slope, mean value and amplitude. The detection process includes three steps. First, the image is filtered by Gaussian filter to smooth the image and filter out the noise. Second, the gradient and direction of each pixel in the image are calculated, the stray response brought by edge detection is eliminated by non-maximum suppression method, and the real and potential edges are identified by double threshold detection method. Third, the isolated weak edge is removed to complete the edge detection.

ii. Harris corner detection

Harris operator is a way to extract the signal feature of images. A fixed window is set and the window moves along the arbitrary directions on the image to be detected, and the change degree of pixel gray level of the window is calculated during the moving process. When the gray level changes greatly, it is considered that there is a corner point in the window. The expression matrix of Harris corner detection is often defined by using the gradient of the horizontal and vertical axis of the local template in the image, as shown in Equation(5 – 10).

$$H = \begin{bmatrix} \sum I_x^2 & \sum I_x I_y \\ \sum I_x I_y & \sum I_y^2 \end{bmatrix} \tag{5 – 10}$$

In the corner detection, there are multiple matching relationships between the corner and the template, which is expressed in Equation(5 – 11).

$$H = \begin{bmatrix} l_2 g^2 \sin^2\theta & l_2 g^2 \sin\theta\cos\theta \\ l_2 g^2 \sin\theta\cos\theta & l_2 g^2 \cos^2\theta + l_1 g^2 \end{bmatrix} \tag{5 – 11}$$

Harris detection has better recognition for corners and edge areas. Because only the first derivative is involved, the calculation process is efficient and easy to operate, and the feature points in the region with rich texture information can be well extracted. It is often used to detect images that are not sensitive to changes in brightness and contrast before and after rotation[114].

iii. Hough transform detection

Hough transform calculates the point-line duality of the image[115]. The given curve of the original image can be detected by mapping the given curve in the original image to a single point in the parameter space and searching the peak value of the parameter space. The detection process is as follows:

(1) Read the original image.

(2) Obtain the source pixel data of the image space.

(3) Using Hough transform method, the pixel coordinates are converted to the curve points, and they are accumulated.

(4) Find the maximum Hough value in the parameter space and set the threshold, and inversely transform the peak value found in the parameter space to the image space.

iv. SIFT

SIFT(scale-invariant feature transform) is an algorithm for detecting the local features of the image. The image features are obtained and feature matching is performed by calculating the feature points in the image[116]. The feature of SIFT is the local feature of the image, which is stable in the detection of viewing angle and noise feature. It is suitable for the fast and accurate matching in the feature database with the advantages of good discrimination and abundant information. Even a few objects can generate a large number of SIFT feature vectors, the optimized SIFT matching algorithm can meet the real-time requirements and can be easily combined with other forms of feature vectors. SIFT calculates and recognizes the local gradient of the image in the neighborhood around the key point in the selected scale, and performs a two-stage recognition process, i. e. , the SIFT features are generated by extracting the feature vector from various images that are independent of size, rotation and brightness, and the SIFT feature vector is matched.

The SIFT calculation process is as follows. Firstly, the extreme value of the scale space is detected, and the scale of all positions in the image is searched. Gaussian differential function is used to identify the point of interest with constant scale and rotation. Secondly, the accurate model is used to locate the location and scale of the identified points of interest. The key points are selected according to the degree of stability, and the local gradient direction of the image is calculated and distributed to one or more directions of each key point position, so that all subsequent operations on the image data can carry out synchronous transformation according to the direction, scale and position of the key points, and the scale-invariant feature is realized.

v. SURF

SURF(speeded-up robust features) is often used for object recognition and 3D reconstruction[117]. The algorithm performance is particularly stable, and its calculation speed is several times faster than the calculation speed of SIFT. It is more stable than SIFT in different image transformation. SURF is an improved algorithm of scale-invariant feature transformation algorithm and can be used to process the objects of matching various images in real time under ideal conditions.

vi. FAST

FAST detection is a method of calculating the value of a pixel in an image. When the difference between a certain pixel in the image and the pixel in its neighborhood is large, the pixel is considered to be a corner. Firstly, select a pixel point in the image and judge whether it is a key point according to its gray value. Secondly, select the appropriate threshold value to calculate the threshold value of the pixels around the point. Finally, the angle is judged according to the calculation results.

The result of using feature detection method to detect normal motion process, runway intrusion and abnormal driving is showed in Figure 5 – 15, Figure 5 – 16 and Figure 5 – 17 respectively.

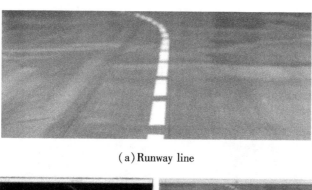

(a) Runway line

(b) Runway line feature detection

Fig. 5 – 15　Runway feature detection

(a) Runway invasion

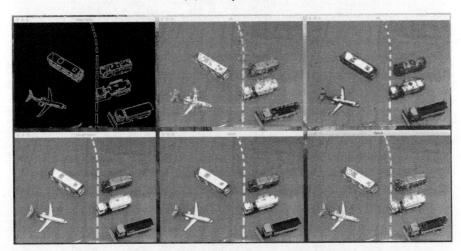

(b) Detection of runway intrusion

Fig. 5 – 16   Runway intrusion feature detection

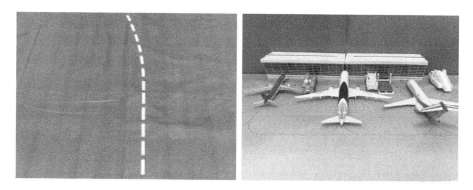

(a) Abnormal driving state

(b) Detection in abnormal driving condition

Fig. 5 – 17   Abnormal driving state feature detection

The recognition of feature points is a part of constraint control. It can effectively distinguish whether the current tractor is in the normal driving. In abnormal motion, the surge of feature points indicates that it is reasonable to use the surge of feature points to judge whether the current motion state is normal.

5.2.2 The environment constraint of FOD

In the motion of the cascade system, the environment constrained motion control was taken to stop the tractor or to avoid the obstacle when the FOD object was detected.

Water fog model detection is an improved algorithm based on image contour detection. Multiple environmental object features in a single image can be quickly and intelligently detected by the model. For a single image, m × n detection points were evenly splashed, and the spline curve, which was constructed with the detection point as the center, was affected by expansion energy, gradient energy and fusion energy simultaneously, as shown in Equation(5 – 12).

$$E_{spray} = \int E_{ext}(f_{(s)}) + E_{gra}(f_{(s)}) + E_{rec}(f_{(s)})\,ds \qquad (5-12)$$

Where,

$E_{ext}(f_{(s)})$ is the expansion energy, which makes the water mist detection area expand to the outside;

$E_{gra}(f_{(s)})$ is the gradient energy, which makes the water mist detection area stop when the gradient change exceeds the threshold;

$E_{rec}(f_{(s)})$ is the fusion energy, when the boundaries of multiple water mist detection areas overlap, the water mist contour stops moving, and the spline function is adjusted to make the detection areas merge.

The expansion energy $E_{ext}(f_{(s)})$ is related to the curvature of the water mist contour, as shown in Equation(5 – 13), and the detection area contour moves in the opposite direction to the center of curvature.

$$E_{ext}(V_{(s)}) = \frac{1}{2}\alpha_{(s)}\left\|\frac{d\,\overline{v}_{(s)}}{ds}\right\| \qquad (5-13)$$

Gradient energy $E_{gra}(f_{(s)})$ is related to the gradient of the contour motion direction, as shown in Equation(5 – 14). This energy is composed of line energy, edge energy, end point energy and their respective weight coefficients. The edge detection weight is much higher than the other two weights in the process of environmental recognition of tractor and aircraft cascade system.

$$E_{gra}(f_{(s)}) = w_{line}E_{line} + w_{edge}E_{edge} + w_{term}E_{term} \qquad (5-14)$$

The water fog model is composed of the energy of each detection point, and the related energy of each water fog model detection point is described in the form of a matrix. The total energy is shown in Equation(5-15).

$$E_{gra}(f_{(s)}) = w_{line}E_{line} + w_{edge}E_{edge} + w_{term}E_{term} \qquad (5-15)$$

The environmental information of the cascade system can be detected and identified quickly and accurately by minimizing , as shown in Equation(5-16).

$$E_{spray}^{*} = \begin{bmatrix} E_{11} & \cdots & E_{1n} \\ \vdots & \ddots & \vdots \\ E_{m1} & \cdots & E_{mn} \end{bmatrix} \qquad (5-16)$$

The detection results of FOD water mist model are shown in Figure 5-18.

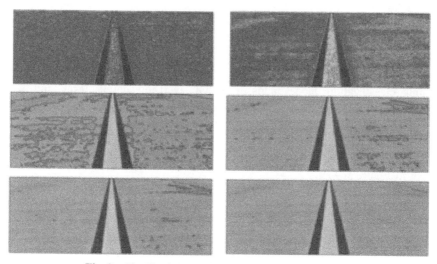

Fig. 5-18　Testing results of FOD water mist model

5.2.3　The environment constraint of runway intrusion

The motion constraint can be effectively controlled in the rapid process such as runway intrusion by using the rapid segmented image intelligent detection during the motion of the tractor. The block algorithm uses a separate convolutional neural network for end-to-end detection. The model consists of a convolutional layer, a pooling layer and a fully connected layer, and the fully connected layer contains two sublayers. As an improved slider search algorithm, the block algorithm transforms the detection problem into an image classification problem[118]. It divides the environmental pictures collected by the cascade system into multiple small squares that do not coincide with each other, then performs convolution operations on these non-coincident small squares to generate a feature map, and uses each feature to set the center point in the small square to predict the target and complete the feature detection of the entire picture[119-120].

In the process of recognition, candidate area detection and object recognition are merged, which can quickly recognize the image category and position information and improve recognition efficiency[121]. In the recognition process, the input image is firstly pooled, and then is partitioned into several small areas, namely, the waiting area. Each block area is allowed to predict the two borders, then the data in the waiting area is put into the convolutional neural network, and finally the target is predicted based on the network model. The block-type algorithm uses a unified framework and end-to-end training, and the whole calculation speed is greatly improved.

i. Block algorithm calculation process

The block algorithm first carries out pre-training on the data set. Its pre-training classification model adds a global average pooling layer and a full connection layer to the convolutional layer in the network, and then adds four randomly initialized convolutional layers and two full connection layers to the convolutional layer obtained from the pre-training to form the whole network, as shown in Figure 5 – 19.

Fig. 5 – 19   Block algorithm network

ii . Block algorithm loss function

The block algorithm defines the loss function as the deviation between the actual output value of the network and the value of the sample tag. It simplifies the target detection to an approximate nonlinear regression problem and uses the mean square error loss function to calculate the weight value of different parts.

The loss function is composed of errors of various structures and extraction areas, including the error of the center point of the border, the error of the height and width of the border, the confidence error when the object is in the border, and the object classification error, as shown in Equation(5 – 17).

$$\lambda_{coord}\sum_{i=1}^{s^2}\sum_{j=0}^{B}A_{ij}^{obj}\left[\left(x_i-x'_i\right)^2+\left(y_i-y'_i\right)^2\right]+\lambda_{coord}\sum_{i=0}^{s^2}\sum_{j=0}^{B}A_{ij}^{obj}$$
$$\left[\left(\sqrt{w_i}-\sqrt{w'_i}\right)^2+\left(\sqrt{h_i}-\sqrt{h'_i}\right)^2\right]+\sum_{i=0}^{s^2}\sum_{j=0}^{B}A_{ij}^{obj}\left(C_i-C'_i\right)^2+$$
$$\sum_{i=0}^{s^2}A_{ij}^{obj}\sum_{c\in classes}\left(p_{i(c)}-p'_{i(c)}\right)^2$$

$$(5-17)$$

iii . Block algorithm network prediction

The block algorithm uses the non-maximum value suppression method to calculate, predicts the characteristic value of each small square, resets the value less than the confidence threshold to zero, determines the category of each small square, and finally outputs the detection results with non-zero confidence value. The detection results of the vehicle intrusion process during runway intrusion are shown in Figure 5 – 20. The upper part of the figure is the runway intrusion process, while the bottom part is the detection results. It can be seen from Figure 5 – 20 that the whole process of runway invasion can be quickly and effectively detected by using the block algorithm.

Fig. 5 – 20   Runway intrusion process and detection results

The classification and detection results of various runway intrusions are shown in Figure 5 – 21. The block algorithm can be used to identify the fire trucks, sewage trucks, gangway trucks and aircraft which enter the camera area accounting for more than 12% of the total area.

Fig. 5 – 21   Detection results of runway intrusion

5.3   Results and discussion

In the process of using machine learning to control the cascade system, the traction process was composed of pulling the aircraft out of the hangar, curve traction and straight traction. The whole traction process was smooth, as shown in Figure 5 – 22.

(a) Pulling the aircraft out of the hangar   (b) Curve traction   (c) Straight traction

Fig. 5 – 22   Traction process

The pushing process was composed of straight pushing, curve pushing and the process of pushing the aircraft into the hangar. The steering gear of the tractor and the aircraft deflection attitude were frequently adjusted in the process, which was a process of continuous data fitting adjustment, as shown in Figure 5 – 23.

(a) Straight pushing　　(b) Curve pushing　　(c) Pushing the aircraft into the hangar

Fig. 5 – 23　Pushing process

The environmental constrained motion control is shown in Figure 5 – 24. After detecting the FOD and the runway intrusion, it controls the cascade system in order to stop the operation.

Fig. 5 – 24　Environmental constrained motion

The experimental results show that the motion control and the environmental restraint control of the cascade system can be performed by using the method of machine learning, and it has good practical effects.

5. 4　Conclusions

Acceleration sensors were used to analyze two motion states of traction and pushing. Linear regression model can be used in traction process, and neural network model with many types of inputs can be used in pushing process.

In combination with supervised learning, neural network and other models, the relationship between each system and model is adjusted. In the simulation experiment, the automatic control of traction and pushing in the course of going straight and turning was verified. The motion state of the cascade system was relatively stable, and the verification degree was relatively high. In particular, the stability control could still be achieved when the runway changed or fluctuated. It shows that machine learning is feasible in engineering application in the control of cascade system.

The motion of aircraft tractor was judged by the methods of feature detection, block rapid detection and water mist model detection. By means of feature detection and intelligent image processing, various targets in the airport were detected and identified. When an abnormality was detected in the process of environmental detection, the motion stop instruction was given to the tractor to realize the stable operation of the cascade system under environmental constraints, and the function of environment perception and collision prediction of the aircraft traction cascade system was performed.

Chapter Six

High-Speed Data Transmission for
the Aircraft Towing Tractor

In the process of aircraft traction, a combination of various sensors is used to obtain the information of the surrounding environment. The acquired state information, attitude information and environment information can be sent to the data room using wireless transmission technology. Data analysis based on machine learning is completed by the powerful computing capacity of the data room, and data transmission is carried out by the network communication mode based on the high-speed link.

6.1 High-Speed communication architecture and environment construction

The high-speed communication link is composed of the automatic control unit of the tractor, the high-speed full-band transmitter and receiver, and the data computing platform, as shown in Figure 6 – 1. The automatic control unit of the tractor obtains the current data of the vehicle and sends the data to the data computing platform through a high-speed link. The data calculation platform quickly analyzes the data and sends control instructions to the tractor, the tractor executes the control instructions.

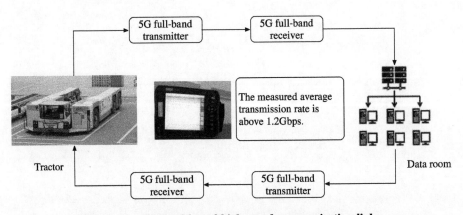

Fig. 6 – 1 Composition of high-speed communication link

In the high-speed link communication process, the interactive communication mode is adopted to enable automatic interactive control between the control equipment of the tractor and the data computing platform[122–123]. The communication process is shown in Figure 6 – 2.

The data connection between the tractor control terminal and the data computing platform is interactive, and the tractor control terminal is automatically

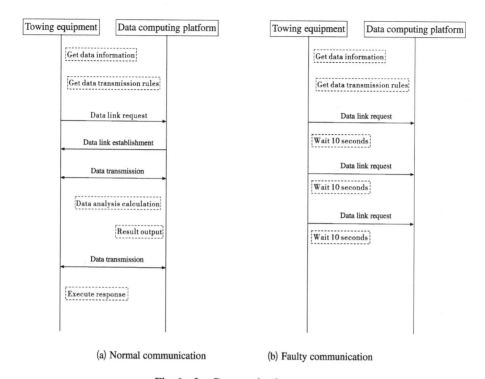

(a) Normal communication   (b) Faulty communication

Fig. 6 – 2   Communication process

connected to the data computing platform after it is turned on. When the connection is established successfully, and there is no normal data packet transmission, the tractor control terminal periodically sends the communication link information of the tractor control terminal to the data computing platform. After receiving the information, the data computing platform will immediately send a reply message to the tractor control terminal. Both the data computing platform and the tractor control terminal can disconnect according to the protocol, and actively determine whether the connection needs to be disconnected.

The conditions for the data computing platform to cut off the data communication link are as follows:

(1) Cut off the communication link according to the protocol.

(2) Have not received data link hold information for a long time.

(3) The same identification code device sends data at the same time.

The conditions for the tractor control terminal to cut off the data communication link are as follows:

(1) Cut off the communication link according to the protocol.

(2) After multiple retransmissions, the link response data is still not received.

(3) Receive garbled code, the main and sub-data packet check is invalid.

The system architecture is shown in Figure 6 – 3. The high-speed wireless network is sliced. The tractor data is sent to the data computing platform through the high speed module with the advantage of low delay and large bandwidth. After data analysis and processing, the platform issues control instructions through the high speed link, and the tractor receives instructions for automatic control.

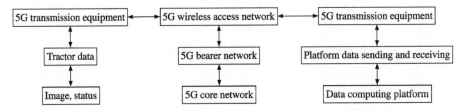

Fig. 6 –3   System architecture

6. 2   High-Speed transmission link design

6.2.1 Transmission link rule design

i . Type of data

The protocol uses network bytes in high-speed network slicing mode to transfer the word and double word. The description information is shown in Table 6 – 1.

Table 6 –1   Description of data type

Data type	Description
BYTE	Unsigned single-byte integer(byte,8 bits)
WORD	Unsigned double-byte integer(word,16 bits)
DWORD	Unsigned four-byte integer(double word,32 bits)
BYTE[n]	n byte
BCD[n]	8421 code,n byte
STRING	UTF-8 code. If there is no data,leave it blank

ii . Rules of data transmission

The data transmission process meets the high-speed wireless network slicing requirements. A standard PDU(protocol data unit)message is transmitted between the tractor and the computing platform through the high-speed wireless network, as shown in Figure 6 – 4.

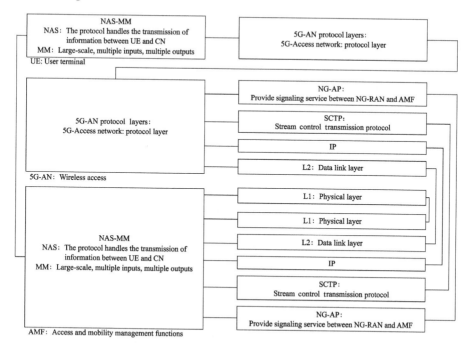

Fig. 6 – 4   Data transmission process

During transmission, the relationship among the PUD message, high-speed terminal and the base station is shown in Figure 6 – 5. The data is under the framework of the basic communication protocols, such as service data adaptation protocol, packet data convergence protocol, and wireless link control self-layer protocol. The data transmission is completed according to the user-defined application communication protocol.

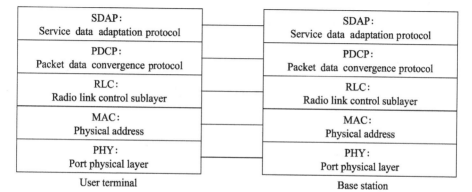

Fig. 6 – 5   Data transmission

iii. Rules of data package

In the transmission process, the data takes the main packet format as the basic transmission unit. The main data packet is composed of five parts: main packet delimiter(head of data Frame), main packet header file, sub-packet, check code and main packet delimiter(end of data Frame), as shown in Figure 6 – 6. The segmentation identification "FF" is inserted in each content.

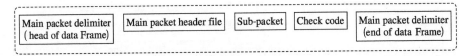

Fig. 6 – 6   Main data packet format

The sub-data package consists of three parts: function name, message name and message content, as shown in Figure 6 – 7.

Fig. 6 – 7   Sub-data packet format

iv. Packet separator and header file rules

(*i*) Packet delimiter

The main packet delimiter is fixed to FF5a. FF5a in the message body needs to be escaped, following these rules:

$$FF5a \rightarrow FF5b \ FF5c$$

$$FF5b \rightarrow FF5b \ FF5d$$

In the sub-data package, the package separator is no longer set separately, and the function name encoding is used to split the packages.

The escape process is as follows:

Sending a message: message encapsulation → compute and fill check code → encryption → escaping.

Receiving a message: escape and restore → decrypt → verify check code → parse the message.

(*ii*) Packet header file

The data packet header file is composed of three parts, i. e., the serial number and attributes of the packet, the identification code of the towing vehicle control terminal and the timestamp, as shown in Figure 6 – 8.

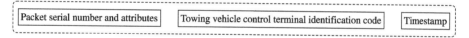

Fig. 6 – 8   Composition of the packet header

Firstly, packet serial number and attributes.

The sending serial number and the encryption signature property in the current packet are recorded. The bit of No. 1 to No. 6 indicates the serial number of the current packet. Add 1 after each message is sent. Reconnect after the connection is disconnected, or the link keeps the data packet synchronization and then clears to zero. The next sending starts from 1. The seventh bit is the data encryption and signature identifier, and the value of 0 is the non-encryption and non-signature. The value of 1 or 2 represents unencrypted first-level and second-level signature respectively, and the value of 3 means the third-level signature and encryption. For example, FF0098663, which means that the current data packet sequence number is 009866, which is encrypted by AES with three-level signature.

Secondly, identification code of the towing vehicle control terminal.

The data computing platform is composed of four parts: the tractor deployment location, vehicle type, order and serial number. Each part is represented by two data bits.

Finally, timestamp.

Greenwich mean time (GMT) is adopted to record the current packet upload time, the data format is year, month, day, hour, minute, second and serial number, each occupying two data bits. For example, FF2020010209190919 indicates that the report time of the 19th data packet is 9: 19: 09 on January 2, 2020.

6.2.2   Sub-packet design

The sub-data packet is the service data packet, which is composed of function name, command name, and instruction content, as shown in Figure 6 – 9.

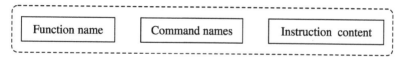

Fig. 6 – 9   Service data packet

The function name is composed of two parts: the basic function and the automatic control function. Its composition and description are shown in Table 6 – 2. The serial number 1 to 10 is the basic function, the serial number 11 to 13 is the automatic control function name, and the reserved serial number is 14 to 15.

Table 6 – 2 Function name of service data package and description

No.	Function name	Function name encoding	Description
1	Link maintenance	FF01	The tractor control terminal sends link maintenance data.
2	Registration binding	FF02	The tractor control terminal sends the registration binding instruction to the data computing platform. The tractor control terminal uploads the DEVICE ID, IMEI, IMSI, and the public key information of the tractor control terminal to the data computing platform.
3	Tractor log-in and log-out	FF03	After the tractor control terminal establishes a connection with the data computing platform, it performs the tractor log-in operation, and actively performs the log-out operation before the tractor control terminal sleeps.
4	Equipment information of tractor control terminal	FF04	The data computing platform requests to obtain the information of the tractor control terminal equipment.
5	Encrypted authentication	FF05	The data computing platform issues a digital key to the tractor control terminal to confirm that the operation is in compliance.
6	Exception reporting	FF06	The tractor control terminal uploads abnormal information to the data computing platform.

(Continued)

No.	Function name	Function name encoding	Description
7	Log reading	FF07	The data computing platform reads the log information of the tractor control terminal.
8	Key level switch	FF08	The data computing platform sends an instruction to switch the key level of the tractor control terminal, which occurs before the remote control instruction is issued, and is optional.
9	Configuration file transfer	FF09	After successfully logging in, the tractor control terminal accesses the data computing platform and compares the version information to check whether it needs to be updated. If necessary, download the configuration file through this interface.
10	Version update	FF0A	The data computing platform remotely upgrades the tractor control terminal.
11	State information reporting	FF0B	Report the state information of the tractor to the data computing platform, including data, images, etc.
12	Remote control	FF0C	Remote control of the tractor based on high-speed data link.
13	Location information	FF0D	Obtain tractor location information.
14	Reserved for basic function	FF11 – FF9F	Design reservations for the basic function expansion of the later system.
15	Reserved for automatic control function	FFA1 – FFFF	Design reservations for the later expansion of automatic control functions.

The instruction name, content and description of the basic function subclass are shown in Table 6 – 3.

Table 6 – 3 Instruction name, content and description of the basic function subclass

No.	Subclass function encoding and name	Instruction name encoding and transmission direction	Instruction content and description (segment code is FF)
1	FF01 Link maintenance	FF01:T→CP	01:Transfer the smallest content
		FF02:CP→T	01:Normal reception 02:Receive error
2	FF02 Registration binding	FF01:T→CP	a ~ A(10 bits):Terminal identification b ~ B(17 bits):Frame identification c ~ C (15 bits): Mobile device international identification code d ~ D (15 bits): International user identification code for mobile devices e ~ E (16 bits): Terminal primary public key
		FF02:CP→T	a(1 bit):Active state b ~ B(16 bits):Primary public key
3	FF03 Tractor log-in and log-out	FF01:T→CP	a ~ A (10 bits): Terminal ID login and logout application
		FF02:CP→T	01:Log in and log out normally 02:Login and logout errors
4	FF04 Equipment information of tractor control terminal	FF01:T→CP	a ~ A(10 bits):Tractor identification code b ~ B (9 bits): Terminal software and hardware version and terminal configuration version of the tractor,3 bits each c(1 bit):Active state
		FF02:CP→T	01:Normal reception 02:Receive error

(Continued)

No.	Subclass function encoding and name	Instruction name encoding and transmission direction	Instruction content and description (segment code is FF)
5	FF05 Encrypted authentication	FF01:T→CP	a ~ A (12 bits): LAN address authentication information b ~ B(16 bits):Communication key
		FF02:CP→T	a(1 bit):Identity verification result b ~ B(12 bits):Communication key
6	FF06 Exception reporting	FF01:T→CP	01:Collision alarm 02:Tractor battery low alarm 03:Low battery alarm of terminal equipment 04:Abnormal driving alarm 05:Abnormal access and control alarm 06:Abnormal location alarm
		FF02:CP→T	01:Normal reception 02:Receive error
7	FF07 Log reading	FF01:T→CP	a ~ A(Log path):Request to report log
		FF02:CP→T	01:Normal reception 02:Receive error
8	FF08 Key level switch	FF01:T→CP	a(1 bit):Request for key level switch, three types
		FF02:CP→T	01:Agree to switch 02:Refuse to switch
9	FF09 Configuration file transfer	FF01:T→CP	a ~ A (Configuration file path):Request to upload configuration file
		FF02:CP→T	01:Agree to transfer 02:Refuse to transmit

(Continued)

No.	Subclass function encoding and name	Instruction name encoding and transmission direction	Instruction content and description (segment code is FF)
10	FF0A Version update	FF01 : T→CP	a ~ A(9 bits) : Request version upgrade, hardware version number, configuration file serial number
		FF02 : CP→T	a ~ A(Update file path) : Agree to upgrade b(1 bit) : Refuse to upgrade

The instruction name, content and description of the automatic control function subclass are shown in Table 6 − 4.

Table 6 − 4 Instruction name, content and description of auto control function subclass

No.	Subclass function encoding and name	Instruction name encoding and transmission direction	Instruction content and description (segment code is FF)
1	FF0B State information report	FF01 : T→CP	a ~ A(Image path) : Upload image information b ~ B(Attitude path) : Upload attitude information of tractor, aircraft and system c ~ C(3bits) : Output power percentage d ~ D(3bits) : Steering angle value e ~ E(3bits) : Speed information f ~ F(5bits) : Mileage information g ~ G (5bits) : Power consumption and fuel consumption information h(1bit) : Gear information i ~ I(7bits) : Door state information j ~ J(7bits) : Window state information k ~ K(Fault file path) : Dashboard display failure l (1bit) : Startup state, including flameout, startup, and operation
		FF02 : CP→T	01 : Normal reception 02 : Receive error

(Continued)

No.	Subclass function encoding and name	Instruction name encoding and transmission direction	Instruction content and description (segment code is FF)
2	FF0C Remote control	FF01 : CP→T	a(1bit) : Steering wheel enable b(1bit) : Drive wheel enable c ~ C(4bits) : Steering wheel steering angle d ~ D(3bits) : Drive wheel output power e(1bit) : Left wheel enable f(1bit) : Right wheel enable g ~ G(3bits) : Left wheel output power h ~ H(3bits) : Right wheel output power i(1bit) : Operation gear shift j(1bit) : Emergency stop k(1bit) : Start authorization l ~ L(6bits) : Door switch m ~ M(6bits) : Window switch n(1bit) : Warning light switch
		FF02 : T→CP	01 : Reception is normal and executed 02 : Receive error
3	FF0D Location information	FF01 : T→CP	a ~ A(20bits) : Latitude and longitude information
		FF02 : CP→T	01 : Normal reception 02 : Receive error

6. 3　Results and discussion

6.3.1　Millimeter wave transmission

The high-speed millimeter wave frequency band was tested, and its transmission rate was stabilized between 1. 2Gbps and 1. 5Gbps. Millimeter wave transmission is capable of large scale data transmission. The results are shown in Figure 6 – 10.

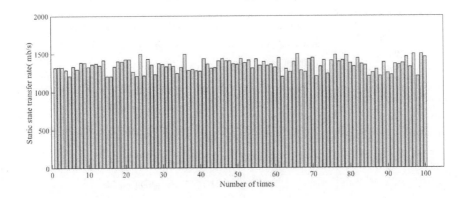

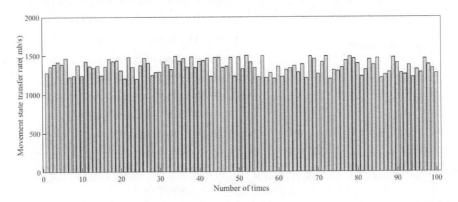

Fig. 6 – 10   Rate test results

6.3.2   Full band communication

Many validation tests were performed based on the high-speed data transmission links, and the test results are shown in Figure 6 – 11. Four channels were set to increase the transmission rate. The first and the third channel were mainly data transmission channels, and the transmission rate was stable at 4. 2 M/s and 4. 9 M/s. The second and the fourth channel were mainly video transmission channels, and the transmission rate was stable at 8. 0 M/s and 8. 6 M/s. The test results show that the transmission rate was stable, and the link bandwidth was stable above 20 M/s, which meets the data transmission requirements.

4. 2M	4. 2M	4. 2M	4. 2M	1. 2 M	4. 2 M
8. 0M	8. 0M	8. 0M	8. 0M	8. 0 M	8. 0 M
N/A	N/A	4. 9M	4. 9M	4. 9 M	4. 9 M
8. 6M	8. 6M	8. 6M	8. 6M	8. 6 M	8. 6M
4. 2M	4. 2M	4. 2M	4. 2M	N/A	4. 2 M
8. 0M	8. 0M	8. 0M	8. 0M	8. 0 M	1. 0 M
4. 9M	4. 9M	4. 9M	4. 9M	4. 9 M	4. 9 M
8. 6M	8. 6M	8. 6M	8. 6M	8. 6 M	8. 6 M
4. 2M	4. 2M	4. 2M	4. 2M	4. 2 M	4. 2 M
4. 2M	4. 2M	4. 2M	4. 2M	4. 2 M	4. 2 M
8. 0M	8. 0M	8. 0M	8. 0M	8. 0 M	8. 0 M
4. 9M	4. 9M	4. 9M	4. 9M	4. 9 M	4. 9 M
8. 6M	8. 6M	8. 6M	8. 6M	8. 6 M	8. 6 M

Fig. 6 – 11   Each channel rate of data transmission link

6.3.3   Response time of control instructions

The response time of control instructions was tested, and the response time was controlled within 6ms in 100 groups of transmission experiments, except for partial data packet loss, as shown in Figure 6 – 12. It shows that the control command delay of using the high-speed link was low, which can meet the functional requirements of the command response to the tractor.

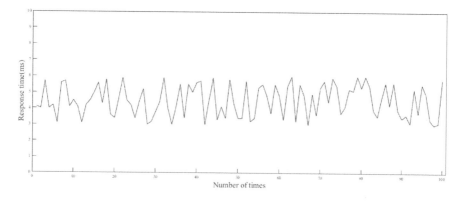

Fig. 6 – 12   Response time test of control instructions

6.4   Conclusions

According to the transmission rules and data packet format rules, the high-speed transmission link was designed, and the high-speed transmission protocol and control commands were determined. Network communication was adopted to connect the data between the tractor control terminal and the data computing platform. Data transmission was realized through high-speed communication rules and communication instructions. During the test, the transmission bandwidth was large, the stability was high, and the delay was low. It has a broad prospect in engineering applications.

The response time of control instructions can be controlled within 6ms by using the high-speed communication network. It can respond to the system instructions quickly, and it can meet the requirements of tractor real-time control in the actual traction process.

Data transmission of aircraft tractors was carried out with high-speed communication networks. The transmission rate of the millimeter wave frequency band was stable from 1.2 Gbps to 1.5 Gbps. It can meet the real-time upload requirements of multi-channel HD video in the actual traction process, and it can realize the rapid unloading of large-scale data.

Chapter Seven

Aircraft Tractor Automatic
Control Simulation

In the research of aircraft tractor automatic control simulation, various subsystems were used to realize the tractor towing and pushing process automatically based on the high-speed data link. The feasibility of each subsystem was verified by measurement and simulation. The algorithm was interpreted, and the subsystems were verified by the simulation system.

7.1 Data collection and annotation

In the simulation experiment, the simulation environment was composed of runway, aircraft simulation module, traction control module, environment module, and high-speed data transmission module, and the simulation ratio was 1 : 30. The data collected include configuration information, control information, runway information, attitude information, environmental information, etc.

7.1.1 Traction platform configuration information

Sampling frequency: 20 Hz.

Maximum number of cycles: 100000 times.

Left turn channel: Channel 1.

Maximum output of PWM for front wheel left steering: 260.

Maximum output of PWM for front wheel right steering: 520.

PWM output of rear wheel forward driving: 200.

PWM output of rear wheel stop driving: 100.

PWM output of rear wheel adjustment motor steering: 0.

Number of samples in a single training session: 128 groups.

Ratio of training set to test set: 0.8.

Single minimum steering angle of steering gear: 1°.

7.1.2 Training data collection and labeling

The training data was acquired by manual operation, and the network operation page was built to control the motion process. The page was composed of five parts: front and rear motion state of the tractor, left and right deflection angle, mode selection, real-time video transmission, and control lever. During the operation, the control interface included five control attitudes: turn left, drive forward left, drive straight, drive forward right, and turn right.

The traction platform sent real-time runway images to the control interface through the network in each training, and the tractor was controlled manually on the control lever. The steering angle and motion mode were displayed on the left side of the page. At the same time, the image information, motor steering angle and motion information were recorded on the server for training data construction. The automatic control platform of the cascade system was controlled manually for many times to obtain the information of runway, motor speed and deflection angle. The collection of training data included multiple processes, such as straight traction, curve traction, straight push, and curve push.

The hold-out method was used to collect data in the tractor towing and pushing process which was completed by manual operation. The acquisition process was periodic sampling. The files collected by the camera were saved in jpg format, and the tractor motor control files were saved in jsonm format. Each process acquired 2000 pictures and control files, and a total of 100000 runway images and control parameter files were collected. The motor control parameter file is composed of the current power preset percentage, deflection angle information, and corresponding image information, as shown in Figure 7 – 1.

record_2599.json	record_5463.json	record_8606.json	record_9082.json
record_2602.json	record_5468.json	record_8612.json	record_9083.json
record_2610.json	record_5469.json	record_8615.json	record_9095.json
record_2612.json	record_5474.json	record_8616.json	record_9101.json
record_2613.json	record_5480.json	record_8622.json	record_9107.json
record_2619.json	record_5483.json	record_8624.json	record_9113.json
record_2621.json	record_5486.json	record_8625.json	record_9115.json
record_2622.json	record_5489.json	record_8633.json	record_9118.json
record_2623.json	record_5492.json	record_8639.json	record_9124.json
record_2630.json	record_5498.json	record_8640.json	record_9136.json
record_2637.json	record_5510.json	record_8643.json	record_9140.json
record_2643.json	record_5514.json	record_8647.json	record_9148.json
record_2644.json	record_5515.json	record_8651.json	record_9149.json
record_2645.json	record_5546.json	record_8660.json	record_9154.json
record_2648.json	record_5547.json	record_8676.json	record_9155.json
record_2661.json	record_5553.json	record_8680.json	record_9159.json
record_2665.json	record_5555.json	record_8691.json	record_9164.json
record_2666.json	record_5557.json	record_8694.json	record_9167.json
record_2668.json	record_5560.json	record_8698.json	record_9168.json
record_2676.json	record_5572.json	record_8699.json	record_9169.json
record_2678.json	record_5575.json	record_8708.json	record_9170.json
record_2680.json	record_5579.json	record_8710.json	record_9176.json
record_2687.json	record_5581.json	record_8717.json	record_9187.json
record_2690.json	record_5582.json	record_8724.json	record_9190.json

(a) Motor control parameters

record_8767.json

(b) Motor control files

⊙ ◯ ◯   record_7912.json — Edited

⊞ < >   record_7912.json ⟩ No Selection

"cam/image_array": "7912_cam-image_array_.jpg"
"power": 0.20,
"angle": 0.571977015904018

⊙ ◯ ◯   record_8311.json — Edited

⊞ < >   record_8311.json ⟩ No Selection

"cam/image_array": "8311_cam-image_array_.jpg"
"power": 0.20,
"angle": 0.1595955984933034

(c) Motor control file information

Fig. 7 – 1   Motor control parameters database (part of the samples)

7.1.3   Environmental restraint control data collection and labeling

Multi-directional image information collection and labeling of a single airport object is shown in Figure $7-2$ (a). The collection included five directions: front, left, right, back and above, and twenty pictures were collected in each direction from $-45°$ to $+45°$. The entire data is shown in Figure $7-2$ (b), including a total of 26 sets of environmental data such as special vehicles, aircraft, and airport buildings. Each set contained 100 original images and 100 labeled images, and a total of 5200 images constituted the environmental constraint data.

(a) Individual vehicle images and labeled information

(b)Environmental information data set

Fig. 7 – 2　Environment data(part of the samples)

In the motion process of the tractor and aircraft cascade system, the surrounding environment involved runways, aircraft, special vehicles, accompanying personnel, towers and buildings, etc. A data set of the main aircraft in China was built, the information includes fuselage height, fuselage length, wing length and maximum take-off mass, etc. The information can be used in the process of runway intrusion in environmental constraints. The accuracy of obstacle avoidance can be improved by increasing the profile parameters of the identified aircraft. The data sets of different aircraft types are shown in Table 7 – 1.

Table 7 – 1　Different aircraft database

Aircraft type	Fuselage height (m)	Fuselage length (m)	Wingspan (m)	Maximum take-off mass (kg)	Number of feature pictures
A318	12. 56	31. 45	34. 10	68000	200
A319	11. 76	33. 84	34. 10	75500	200
A320	11. 76	37. 57	34. 10	77000	200
A321	11. 76	44. 51	34. 10	93500	200
A330-200	17. 40	58. 80	60. 30	230000	200

(Continued)

Aircraft type	Fuselage height (m)	Fuselage length (m)	Wingspan (m)	Maximum take-off mass (kg)	Number of feature pictures
A330-300	16. 85	63. 60	60. 30	230000	200
A330-200F	16. 90	58. 80	60. 30	230000	200
A340-200	16. 70	59. 39	60. 30	275000	200
A340-300	16. 85	63. 60	60. 30	276500	200
A340-500/ 500 HGW	17. 10	67. 90	63. 45	372000/ 380000	200
A340-600/ 600 HGW	17. 30	75. 30	63. 45	368000/ 380000	200
A350-800	17. 05	60. 54	64. 75	248000	200
A350-900	17. 05	66. 80	64. 75	268000	200
A350-1000	17. 08	73. 78	64. 75	308000	200
A380	24. 09	72. 75	79. 75	575000	200
B707-120B	12. 70	44. 22	39. 88	117000	200
B707-320/ 420	12. 85	46. 61	43. 41	141700	200
B707-320C	12. 80	46. 61	44. 42	151500	200
B727	10. 36	46. 69	32. 92	72570	200
B737-300	4. 01	33. 40	28. 90	62000	200
B737-400	4. 11	36. 40	28. 90	68050	200
B737-700	12. 50	33. 60	34. 30/35. 70 (with winglet)	79010	200
B737-800	12. 50	39. 50	34. 40/35. 79 (with winglet)	78245	200
B747-100	19. 30	70. 60	59. 60	333390	200
B747-200B	19. 30	70. 60	59. 60	377842	200
B747-300	19. 30	70. 60	59. 60	377842	200
B747-400	19. 40	70. 60	64. 40	396890	200
B747-400ER	19. 40	70. 60	64. 40	412775	200

(Continued)

Aircraft type	Fuselage height (m)	Fuselage length (m)	Wingspan (m)	Maximum take-off mass (kg)	Number of feature pictures
B747-8	19. 30	76. 40	68. 50	439985	200
B757-200	13. 56	47. 33	38. 05	115650	200
B757-200F	13. 56	47. 33	38. 05	116650	200
B757-300	13. 56	54. 41	38. 05	122470	200
B767-200	5. 41	48. 50	47. 57	142880	200
B767-200ER	5. 41	48. 50	47. 57	179170	200
B767-300	5. 41	54. 90	47. 57	158760	200
B767-300ER	5. 41	54. 90	47. 57	186880	200
B767-300F	5. 41	54. 90	47. 57	186880	200
B767-400ER	5. 41	61. 40	51. 82	204120	200
B777-200	6. 19	63. 70	60. 90	247210	200
B777-200ER	6. 19	63. 70	60. 90	297560	200
B777-200LR	6. 19	63. 70	64. 80	347450	200
B777 F	6. 19	63. 70	64. 80	347450	200
B777-300	6. 19	73. 90	60. 90	299370	200
B777-300ER	6. 19	73. 90	64. 80	351534	200
B787-8	17. 00	56. 69	60. 17	227930	200
B787-9	17. 00	63. 00	60. 17	247000	200
B787-10	17. 00	68. 00	60. 17	N/A	200
Dornier 328	7. 05	21. 23	20. 98	15660	200
CRJ-200 ER	6. 22	26. 77	21. 21	23100	200
CRJ-700 ER	7. 57	32. 51	23. 24	34000	200
MD 90	9. 33	46. 51	32. 87	72803	200
E145	7. 10	27. 96	20. 53	17500	200
ATR72	7. 65	27. 20	27. 05	22500	200
CESSNA208	4. 27	12. 70	15. 90	3969	200
Hawker 800	5. 26	15. 37	14. 32	11340	200
Y-7	8. 55	23. 70	29. 20	21800	200

(Continued)

Aircraft type	Fuselage height (m)	Fuselage length (m)	Wingspan (m)	Maximum take-off mass (kg)	Number of feature pictures
Y-8	11.98	36.80	42.80	81000	200
BAe146-100	8.61	26.16	26.21	38100	200
BAe146-200	8.61	28.55	26.21	42185	200
BAe146-300	8.61	30.99	26.21	24878	200
CL-600	6.30	20.85	19.61	19550	200
TU154 B-2	11.40	48.00	37.55	98000	200
TU154 M	11.40	48.00	37.55	102000	200
YAK-42	9.83	36.38	34.88	56500	200
ARJ21	8.44	33.46	27.28	43500	200

In the research, the feature pictures of 39 types of aircraft were collected and classified in the way of folder name, which were used to train the machine learning network to recognize the real aircraft model. The higher dimensional environment perception ability was constructed according to the contour parameter information. The data set, with about 7800 images, was composed of 39 groups of data sets, and each of them contained about 200 images of a single model. Part of the data is shown in Figure7-3. Figure 7 – 3 (a) is a set of 200 feature images of a single model, and Figure 7 –3(b) is the information of a single feature image.

(a) Single model feature images

(b) Single feature image information

Fig. 7 – 3　Aircraft features(part of the samples)

A total of 360 images of 18 types of tractor were collected in the research, as shown in Figure 7 –4(a) , which were used to train the machine learning network to recognize the tractor in a real environment. The image annotation information is shown in Figure 7 –4(b).

(a) Tractor database in real environment

(b) Image annotation information

Fig. 7 – 4　Tractor feature atlas(part of the samples)

7. 2　Model training

7.2.1　Model structure

The machine learning network includes five parts: input, feature extraction, model construction, network layer and automatic control output.

ⅰ. Input

The image information, motor motion state information and cascade system posture information were obtained during manual operation.

ⅱ. Feature extraction

Features are extracted from the input data by many different methods in order to identify the curve and straight motion information, towing and pushing state information, and environmental information.

ⅲ. Model building

The machine learning model of aircraft traction cascade system was constructed. The model was combined with the supervised learning model, multi-input neural network model and back propagation model. The cascade system stability maintenance model was introduced to smooth the large-scale motor output waveform jump caused by manual operation in the control process.

ⅳ. Network configuration

Network configuration is used to analyze, identify and calculate the model, including data preprocessing, data analysis, data manipulation and main layers. Data preprocessing includes text preprocessing, picture preprocessing and sequence preprocessing. Data analysis includes multi-type input neural network processing, back propagation processing, gradient analysis processing and support vector machine analysis processing. Network configuration involves many algorithms, some of which are redundant design. Data operations include data flow executor, matrix operation, Matmul, Bayesian operation, Queue, ReLU, parameter update, Softmax, cross entropy, etc. The main layers of the network include common layer, convolutional layer, activation layer, local link layer, pooling layer, loop layer and series layer.

ⅴ. Automatic control output

Automatic control output realizes the corresponding automatic control output of the aircraft traction cascade system under the condition of different runways and the main features are not changed. The collected data is divided into training set and test set after data collection. And then the training data features are extracted; the machine writing model is constructed; the machine learning network configuration is set. Finally, the real-time traction process is output.

7.2.2   Model training process

The image information was input to the machine learning network for network training and generating automatic control model according to the steps of model selection, network layer construction, compilation, training, and generation verification. Figure $7-5(a)$, Figure $7-5(b)$ and Figure $7-5(c)$ show the loss function value at 0. 1254, 0. 1162 and 0. 1090, respectively. The model loss function decreasing and converging indicates that the model fitting is improved continuously in the training.

```
W tensorflow/core/framework/allocator.cc:101 Allocation of 16367616 exceeds 10% of system memory.
W tensorflow/core/framework/allocator.cc:101 Allocation of 11612160 exceeds 10% of system memory.
W tensorflow/core/framework/allocator.cc:101 Allocation of 11612160 exceeds 10% of system memory.
W tensorflow/core/framework/allocator.cc:101 Allocation of 16367616 exceeds 10% of system memory.
W tensorflow/core/framework/allocator.cc:101 Allocation of 22848000 exceeds 10% of system memory.
W tensorflow/core/framework/allocator.cc:101 Allocation of 22848000 exceeds 10% of system memory.
W tensorflow/core/framework/allocator.cc:101 Allocation of 55590912 exceeds 10% of system memory.
W tensorflow/core/framework/allocator.cc:101 Allocation of 31168800 exceeds 10% of system memory.
W tensorflow/core/framework/allocator.cc:101 Allocation of 24429600 exceeds 10% of system memory.
ETA:2:14-loss:0.1254-angle_out_loss:0.2087-throttle_out_loss:0.04212020
```

(a) Loss value 0. 1254

```
W tensorflow/core/framework/allocator.cc:101 Allocation of 16367616 exceeds 10% of system memory.
W tensorflow/core/framework/allocator.cc:101 Allocation of 11612160 exceeds 10% of system memory.
W tensorflow/core/framework/allocator.cc:101 Allocation of 11612160 exceeds 10% of system memory.
W tensorflow/core/framework/allocator.cc:101 Allocation of 16367616 exceeds 10% of system memory.
W tensorflow/core/framework/allocator.cc:101 Allocation of 22848000 exceeds 10% of system memory.
W tensorflow/core/framework/allocator.cc:101 Allocation of 22848000 exceeds 10% of system memory.
W tensorflow/core/framework/allocator.cc:101 Allocation of 55590912 exceeds 10% of system memory.
W tensorflow/core/framework/allocator.cc:101 Allocation of 31168800 exceeds 10% of system memory.
W tensorflow/core/framework/allocator.cc:101 Allocation of 24429600 exceeds 10% of system memory.
ETA:3:21-loss:0.1162-angle_out_loss:0.1984-throttle_out_loss:0.03762020
```

(b) Loss value 0. 1162

W tensorflow/core/framework/allocator.cc:101 Allocation of 16367616 exceeds 10% of system memory.
W tensorflow/core/framework/allocator.cc:101 Allocation of 11612160 exceeds 10% of system memory.
W tensorflow/core/framework/allocator.cc:101 Allocation of 11612160 exceeds 10% of system memory.
W tensorflow/core/framework/allocator.cc:101 Allocation of 16367616 exceeds 10% of system memory.
W tensorflow/core/framework/allocator.cc:101 Allocation of 22848000 exceeds 10% of system memory.
W tensorflow/core/framework/allocator.cc:101 Allocation of 22848000 exceeds 10% of system memory.
W tensorflow/core/framework/allocator.cc:101 Allocation of 55590912 exceeds 10% of system memory.
W tensorflow/core/framework/allocator.cc:101 Allocation of 31168800 exceeds 10% of system memory.
W tensorflow/core/framework/allocator.cc:101 Allocation of 24429600 exceeds 10% of system memory.
ETA:15s-loss:0.1090-angle_out_loss:0.1849-throttle_out_loss:0.03322020

(c) Loss value 0. 1090

Fig. 7 – 5   Network training process

Table 7 – 2 shows the model training group and training time. The hardware configuration is as follows:

Processor:2. 3 GHz Intel Core i7;

Memory:16 GB 1600 MHz DDR3;

Graphics card:Intel HD Graphics 4000,1536 MB;

Hard disk:1 TB solid state SATA driver.

The training time is positively correlated with the number of training samples. The training time was significantly increased with the increasing of training data, and the fitting was also enhanced.

Table 7 – 2   Model training groups and time

Training group	Training time(minutes)
1	945
2	819
3	714
4	1209
5	1005

7. 3   Model verification

7.3.1 Traction process

The traction experiments are shown in Figure 7 – 6.

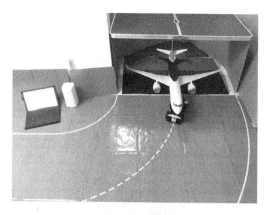

(a) Out of the hangar

(b) Curve traction

(c) Straight traction

Fig. 7 – 6   Traction experiments

The acceleration, angular velocity and angle changes of the simulated motion of the tractor and the aircraft in the traction process are shown in Figure 7 – 7.

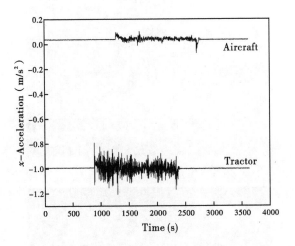

(a – 1) Acceleration in x direction

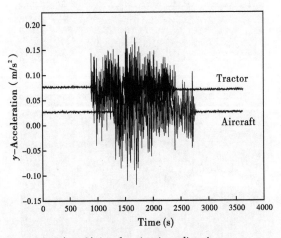

(a – 2) Acceleration in y direction

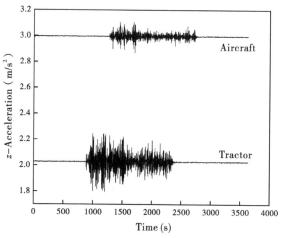

(a - 3) Acceleration in z direction

(a) Acceleration

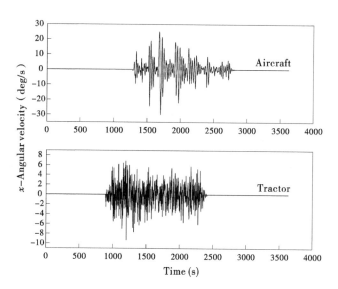

(b - 1) Angular velocity in x direction

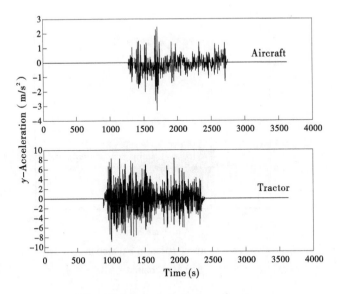

(b-2) Angular velocity in y direction

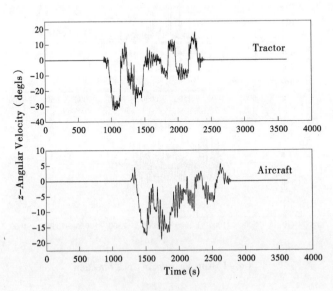

(b-3) Angular velocity in z direction

(b) Angular velocity

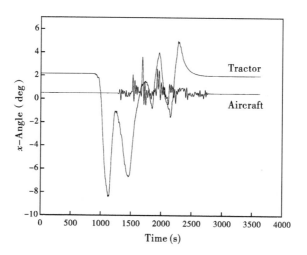

(c − 1) Angle in *x* direction

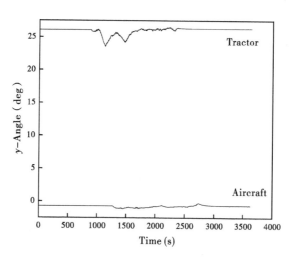

(c − 2) Angle in *y* direction

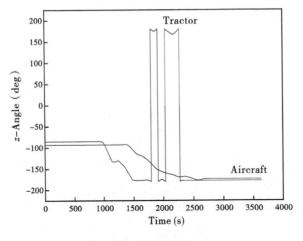

(c – 3) Angle in z direction

(c) Angle

Fig. 7 – 7   Attitude information in the traction process

7.3.2   Pushing process

The process of pushing experiments are shown in Figure 7 – 8.

(a) Straight pushing

(b) Curve pushing

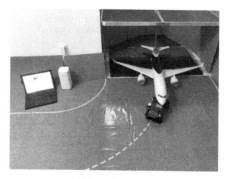

(c) Into the hangar

Fig. 7 – 8   Process of pushing experiments

The acceleration, angular velocity and angle changes of the simulated motion of the tractor and aircraft in the process of pushing are shown in Figure 7 – 9.

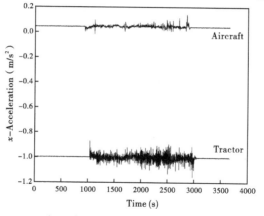

(a – 1) Accelerationin in x direction

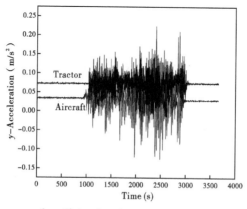

(a – 2) Accelerationin in y direction

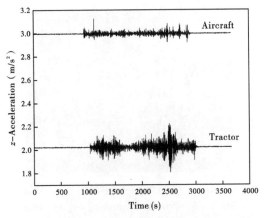

(a – 3) Accelerationin in z direction

(a) Acceleration

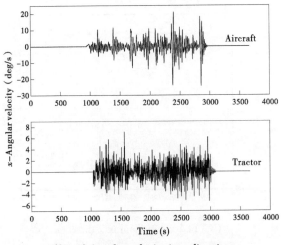

(b – 1) Angular velocity in x direction

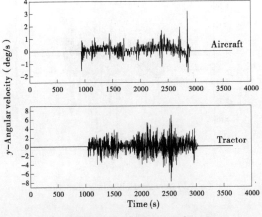

(b – 2) Angular velocity in y direction

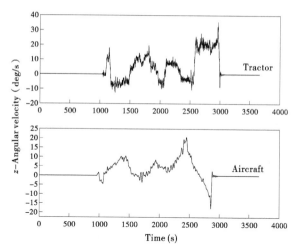

(b-3) Angular velocity in z direction

(b) Angular velocity

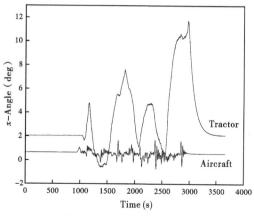

(c-1) Angle in x direction

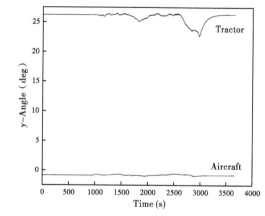

(c-2) Angle in y direction

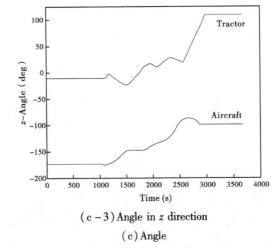

(c − 3) Angle in z direction

(c) Angle

Fig. 7 − 9   Attitude information in the pushing process

7.3.3   Environment constrained motion control

The airport runway was detected during traction. The intrusion detection with different proportions and different objects is shown in Figure 7 − 10.

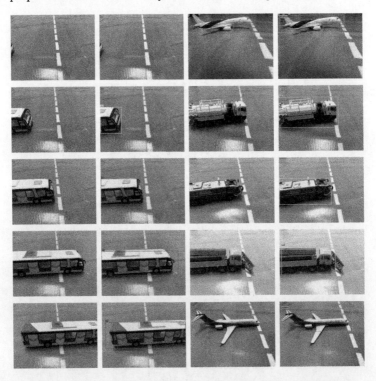

Fig. 7 − 10   Block algorithm detection(part of the samples)

7.3.4   Results of high-speed data link test

The data transmission based on the high-speed data link was stable in the motion process of the cascade system. The results of high-speed data link transmission test are shown in Table 7 – 3.

Table 7 – 3   Results of high-speed data link transmission test

Channel	Main transmission	Main verification	Sub-transmission	Sub-verification
1	1712380	23288	1696358	39310
1	1728499	25069	1711434	42134
1	1756236	26152	1738503	43885
1	1770911	26205	1753148	43968
1	1770911	26205	1753148	43968
1	1811265	26279	1793452	44092
1	1828011	26493	1810121	44383
1	1843983	26497	1826091	44389
1	1860375	26497	1842483	44389
1	1878723	26505	1860819	44409
1	1893965	26655	1875946	44647
1	1915516	26676	1897482	44710
1	1928268	26680	1910231	44717
1	1941456	26716	1943396	44776
1	1948390	26714	1930332	44772
1	1975138	26726	1957074	44790
1	2069887	26889	2051753	45023
1	2084093	26891	2065955	45029
1	2097870	26942	2079711	45101
1	2110527	26945	2092365	45107
1	2136871	26945	2118709	45107
1	2154711	26945	2136549	45107
1	2172443	26949	2454277	45115
1	2185099	26949	2166933	45115
2	3757170	19122	3757170	19122

(Continued)

Channel	Main transmission	Main verification	Sub-transmission	Sub-verification
2	3795175	19124	3795175	19124
2	3847370	19147	3847370	19147
2	3922812	20277	3922812	20277
2	3957368	20416	3957368	20416
2	3989894	20416	3989894	20416
2	4078573	20531	4078573	20531
2	4115750	20533	4115750	20533
2	4150884	20535	4150884	20535
2	4186938	20535	4186938	20535
2	4227286	20570	4227286	20570
2	4308489	20574	4308489	20574
2	4261041	20574	4261041	20574
2	4336551	20583	4336551	20583
2	4380826	20714	4380826	20714
2	4409551	20780	4409551	20780
2	4439690	20782	4439690	20782
2	4648194	20997	4648194	20997
2	4679499	20999	4679449	20999
2	4709802	21129	4709802	21129
2	4772315	21139	4772315	21139
2	4834817	21142	4834817	21142
2	4873834	21149	4873834	21149
2	4901663	21157	4901663	21157
3	4898420	406	4898155	671
3	4948558	551	4948429	860
3	5020882	551	5020537	860
3	5055460	551	5055151	860
3	5055460	551	5055151	860
3	5150428	551	5150119	860
3	5190127	551	5189818	860

(Continued)

Channel	Main transmission	Main verification	Sub-transmission	Sub-verification
3	5227630	551	5227321	860
3	5266096	569	5265771	894
3	5309197	569	5308872	894
3	5345227	575	5344898	904
3	5395879	575	5395550	904
3	5425820	577	5425489	908
3	5473061	577	5472730	908
3	5503742	577	5503411	908
3	5535880	587	5535543	924
3	5758384	635	57580116	1003
3	5791747	635	5791379	1003
3	5824129	635	5823761	1003
3	5853535	625	5853470	1003
3	5915696	635	5915327	1003
3	5957564	643	5957191	1016
3	5999207	643	5998834	1016
3	6028916	643	6028543	1016
4	9619164	1296	9603351	17109
4	9700142	1298	9684321	17119
4	9811787	1353	9795913	17227
4	9982525	1355	9966649	17231
4	10057145	1355	10041269	17231
4	10126425	1355	10110549	17231
4	10316665	1355	10600789	17231
4	10396165	1355	10380283	17237
4	10471283	1357	10455399	17241
4	10634682	1378	10618782	17278
4	10548342	1387	10532442	17278
4	10706844	1396	10690934	17306
4	10808304	1396	10792394	17306

(Continued)

Channel	Main transmission	Main verification	Sub-transmission	Sub-verification
4	10868300	1400	10852388	17312
4	10962940	1400	10947028	17312
4	11024400	1400	11008488	17312
4	11088750	1410	11072827	17333
4	11534480	1460	11518552	17418
4	11601256	1504	11584643	18117
4	11666078	1562	11648057	19583
4	11849083	1977	11823679	27381
4	11932921	1999	119074933	27427
4	12016291	2029	11990597	27723
4	12075517	2323	12044265	33575

7.4 Conclusions

The database was established by collecting runway image, turning angle, motor speed and environmental characteristic information. The training set and test set were obtained by classifying the database. The aircraft tractor motion control simulation was realized by model training and test verification.

The movement of aircraft and tractors was controlled by the feature detection, FOD detection, runway intrusion detection, distance detection, etc. Stop action can be taken in time to ensure safety in abnormal conditions.

The database of different types of aircraft was established to realize the environmental information detection and recognition in the aircraft tractor motion process. The data set was composed of ground clearance, aircraft length, wingspan, take-off weight, and feature atlas. The aircraft data set was used to realize the environmental perception function of the tractor and aircraft cascade system by comparing the identified aircraft type.

The automatic control and data transmission of the aircraft tractor were completed in the simulation experiment by using machine learning and high-speed data link respectively.

Chapter Eight

Summary and Prospect

8.1 Summary

In recent years, many new requirements have been put forward for China's large passenger aircraft development, test flight and operation with the rapid improvement of China's airport construction level and the continuous expansion of the construction scale. This book takes the aircraft tractor automatic control system as the research object, applies machine learning, high-speed data link and intelligent image processing to the aircraft tractor automatic control simulation, and provides technical support for further development of new aircraft tractors in China's national conditions. The study on the aircraft tractor automatic control system is of great significance to improve the airport's and tractor's safety and the efficient operation.

This book reviews the current researches of the aircraft tractor automatic control system, machine learning, airport environment detection and recognition, pattern recognition and intelligent image processing. The method of machine learning was used to establish the traction and push models in order to realize the aircraft tractor automatic control, and the communication control protocols based on the high-speed transmission links were designed. In the simulation control process, the stability of the aircraft traction cascade system, aircraft tractor movement environment recognition and motion state control were studied by the machine learning technology, intelligent image processing, data set construction and high-speed data transmission. The aircraft tractor cascade system was studied by simulation, and the automatic control of the aircraft tractor was realized. The research results are summarized as follows:

The multi-scale simulation environment was established for the aircraft traction cascade system. The traction process and the push process of the cascade system were analyzed. The system function, simulation parameters and the system architecture were designed.

The stable control of the aircraft tractor cascade system could be realized by PID(proportional, integral, derivative) control, linear quadratic regulator control, sliding film variable structure control and fuzzy control. The motor speed and angle waveforms in stable holding, straight driving and initial angle control were studied in the stability control experiments. The motion stability of the cascade system was high in the numerical fitting of the angular acceleration sensor.

The automatic control model of multi-type input neural network was designed with image information and motion sensor to realize the automatic control in the process of traction and pushing. The image analysis and detection methods were used to constrain the motion process in abnormal conditions, and the functions of environmental perception and collision prediction of the aircraft traction cascade system were constructed and verified by simulation.

The methods of intelligent image processing, feature and contour detection were used to identify the features and contours of the surrounding environment in the automatic control process of the aircraft traction cascade system. The fast intelligent image detection was used to realize the object detection and classification of single image.

The interactive mode was adopted to realize the high-speed data transmission and rapid response of the tractor control terminal based on the high-speed data link, which can meet the real-time control requirements in the actual traction process.

The data connection between the tractor control terminal and the data computing platform was carried out by TCP, and the data transmission was realized with 5G communication rules and instructions. The 5G communication network could achieve the data bandwidth of 1. 2 Gbps, and it reduced the response time of the control instructions to less than 10ms and quickly responded to the system instructions, so as to meet the real-time control requirements.

The database of different types of aircraft was established. It is composed of the ground clearance, the length of the fuselage, wingspan, take-off weight, and feature atlas. The runway image, turning angle, motor speed and environmental characteristic information were collected, and the training data set and the test data set were obtained by classifying the database. The data of the aircraft driving and entering hangar by tractor was taken as the training data, the neural network and the semi-supervised learning were used to train the system. The decision tree was used as an auxiliary judgment tool to obtain the more complete machine learning model.

8. 2   Prospect

With the development of artificial intelligence [124 – 125] , aircraft tractors will develop in the direction of more efficiency, energy saving, emission reduction and intelligence in order to realize the intelligent operation and diagnosis of the aircraft traction cascade system and to meet the comprehensive guarantee of various working conditions of the airport. The core elements of artificial intelligence are composed of data, algorithms and computing capabilities. Data is the foundation, algorithms are the core, and computing power is the guarantee. Therefore, the future research should focus on the test flight and operation requirements of China's large passenger aircraft based on the existing foundation. The intelligent automatic control of the aircraft tractor should be further studied in the following aspects.

8.2.1 Database construction

Data is the basis of all calculations and applications. The location of data creation includes core, edge, and endpoint. The research has preliminarily established the relevant data and feature atlas of the main aircraft models. In the future research, data information such as aircraft hangar, runway, tower and personnel should be established in combination with the actual test flight of large passenger aircraft, so as to improve the test flight command center and cloud data. The intelligent operation of aircraft traction cascade system could be realized by high-speed transmission data link.

8.2.2 Research and application of machine learning algorithms

Machine learning is the fundamental way to make machines smart. Traditional machine learning algorithms are mainly decision trees, random forests, artificial neural networks, Bayesian learning, support vector machines, association rules, deep learning [126], etc. At present, machine learning has become a common research hotspot in the field of artificial intelligence and pattern recognition, and its theories and methods are widely used to solve the complex problems in the engineering applications and science. The field of machine learning research can be divided into two categories. The first is the traditional machine learning research, mainly studying learning mechanisms, focusing on exploring the learning mechanism of simulated humans. The second is the research of machine learning in the environment of big data, studying how to use information effectively, focusing on obtaining effective and understandable knowledge hidden from huge amounts of data. The efficient acquisition of knowledge through machine learning has gradually become the main driving force for the development of machine learning with the increasing demand for data analysis in various industries in the era of big data. It is the main research direction for the machine learning to analyze the complex and diverse data deeply and to use the information more efficiently in the current big data environment. Therefore, machine learning is increasingly moving in the direction of intelligent data analysis and it has become an important source of intelligent data analysis technology. The future study should focus on the research of aircraft tractor machine learning in the big data environment, and should be combined with the actual environment of the airport to study the machine learning algorithm of the intelligent aircraft traction cascade system to meet the needs of big data processing based on the existing research.

8.2.3   Automatic and driverless control of airport special vehicles

The automatic control of airport special vehicles will become the main technology in the future. Special vehicles will be designed as highly integrated intelligent products such as machinery, electricity, hydraulic and gas, with the intelligence getting higher and higher. The core idea of automatic control of special vehicles is to change the existing manual operation mode to realize the automatic control and strategy planning based on machine learning and high-speed data link, which can improve operation efficiency.

In terms of unmanned driving, the unmanned aircraft tractor will tow the aircraft to the air bridge after landing, and also takes the aircraft directly onto the runway when it leaves the parking apron. This is of great significance for energy conservation and emission reduction, reducing failure rate and improving the efficiency of vehicle utilization.

References

[1] 周志华. 机器学习 [M]. 北京：清华大学出版社，2019.

[2] SHTERN Y I, EGOROV V A, KARAVAEV I S, et al. Method and hardware software facilities for studying wireless communication channels in smart devices for data measurement and transmission [J]. *Measurement Techniques*, 2016, 59（2）：183 – 187.

[3] SHU Y, ZHU F. Green communication mobile convergence mechanism for computing self-offloading in 5G networks [J]. *Peer-to-Peer Networking and Applications*, 2019, 12（6）：1511 – 1518.

[4] MA Z, ZHANG Z, DING Z, et al. Key techniques for 5G wireless communications：network architecture, physical layer and MAC layer perspectives [J]. *Science China, Information Sciences*, 2015, 58（4）：1 – 20.

[5] 闫眠，赵立军，姜继海，等. 飞机牵引车结构及动力传动系统的发展 [J]. 液压与气动，2009（12）：1 – 4.

[6] ROTH D, JACOBS G, PIETRZYK T, et al. Decentralized compact hydraulic power supply by high speed components [J]. *ATZ Heavy Duty Worldwide*, 2019, 12（1）：66 – 71.

[7] WEICH C, BAYER D, PUCKMAYR D. Challenges of functional safety in tractor development [J]. *ATZ Offhighway Worldwide*, 2017, 10（4）：56 – 59.

[8] ZHANG R, YU X, HU Y, et al. Active control of hydraulic oil contamination to extend the service life of aviation hydraulic system [J]. *The International Journal of Advanced Manufacturing Technology*, 2018, 96（5 – 8）：1693 – 1704.

[9] YANG H, PAN M. Engineering research in fluid power：a review [J]. *Journal of Zhejiang University, Science A*, 2015, 16（6）：427 – 442.

[10] 朱贺，王立文，罗心悦. 空气悬架无杆飞机牵引车动力学仿真分析 [J]. 机床与液压，2018, 46（13）：144 – 147.

[11] 刘成鑫，刘晖，沈圳. 基于 Adams 飞机两种牵引方式控制仿真分析 [J]. 工业控制计算机，2018, 31（1）：77 – 81.

[12] WANG N, LIU H, YANG W. Path tracking control of a tractor aircraft system [J]. *Journal of Marine Science and Application*, 2012, 11（4）：512 – 517.

[13] ALONSO T D, MORA C F. Aircraft ground operations：steps towards automation [J]. *CEAS Aeronautical Journal*, 2019, 10（3）：965 – 974.

[14] HASAN Y J, SACHS F, DAUER J C. Preliminary design study for a future unmanned cargo aircraft configuration [J]. *CEAS Aeronautical Journal*, 2018, 9 (4): 571 –586.

[15] EGOROVA O A, DARSHT Y A, KUZNETSOVA S V. Modeling of automatic control mechanism for the hydraulic transmission of a transportation robot under nonlinear motion characteristics [J]. *Automation and Remote Control*, 2018, 79 (4): 768 –773.

[16] HASAN M E, GHOSHAL S K, DASGUPTA K, et al. Dynamic analysis and estimator design of a hydraulic drive system [J]. *Journal of the Brazilian Society of Mechanical Sciences and Engineering*, 2017, 39 (4): 1097 – 1108.

[17] KIM I S, LEE W K, HONG Y D. Simple global path planning algorithm using a ray casting and tracking method [J]. *Journal of Intelligent & Robotic Systems*, 2018, 90 (1 –2): 101 –111.

[18] ARAJI A S. Development of kinematic path tracking controller design for real mobile robot via back stepping slice genetic robust algorithm technique [J]. *Arabian Journal for Science and Engineering*, 2014, 39 (12): 8825 –8835.

[19] BIJJAHALLI S, RAMASAMY S, SABATINI R. A novel vehicle based GNSS integrity augmentation system for autonomous airport surface operations [J]. *Journal of Intelligent & Robotic Systems*, 2017, 87 (2): 379 –403.

[20] 陈嘉伟, 王金栋, 曲兴华, 等. 光频梳频域干涉测距主要参数分析及一种改进的数据处理方法 [J]. 物理学报, 2019, 68 (19): 39 –49.

[21] 周宣赤, 张孝兵, 张宏峰, 等. 基于 PSO – SVM 的车辆防碰撞预警模型研究 [J]. 控制工程, 2018, 25 (1): 62 –70.

[22] DAMATO E, MATTEI M, NOTARO I. Distributed reactive model predictive control for collision avoidance of unmanned aerial vehicles in civil airspace [J]. *Journal of Intelligent & Robotic Systems*, 2019 (6): 1 –19.

[23] RAMASAMY S, SABATINI R, GARDI A. A unified analytical framework for aircraft separation assurance and UAS sense and avoid [J]. *Journal of Intelligent & Robotic Systems*, 2018, 91 (3 –4): 735 –754.

[24] BAUER P, HIBA A, BOKOR J, et al. Three dimensional intruder closest point of approach estimation based on monocular image parameters in aircraft sense and avoid [J]. *Journal of Intelligent & Robotic Systems*, 2019, 93 (1 –2): 261 –276.

[25] 郭艳颖. 基于视觉的飞机泊位自动引导关键技术研究 [D]. 南京: 南京航空航天大学, 2012.

[26] ITOH E, WICKRAMASINGHE N K, HIRABAYASHI H, et al. Feasibility study on fixed flight path angle descent for wide body passenger aircraft [J]. *CEAS Aeronautical Journal*, 2019, 10 (2): 589 –612.

[27] HUIKARI J, AVRUTIN E, RYVKIN B, et al. High energy subnanosecond optical pulse generation with a semiconductor laser diode for pulsed TOF laser ranging utilizing the single photon detection approach [J]. *Optical Review*, 2016, 23 (3): 522 –528.

［28］ BAILEY K, CURRAN K. An evaluation of image based steganography methods using visual inspection and automated detection techniques ［J］. *Multimedia Tools and Applications*, 2006, 31 (3): 327 – 331.

［29］ 李亚玲. 飞机泊位引导系统管理与监控软件设计 ［D］. 成都：电子科技大学, 2015.

［30］ 张恩迪. 机场飞机入坞管理系统设计 ［D］. 成都：电子科技大学, 2014.

［31］ 马创. 基于光学的飞机泊位指示系统设计 ［D］. 天津：中国民航大学, 2011.

［32］ ELIBOL A, CHONG N Y, SHIM H, *et al*. Mismatched image identification using histogram of loop closure error for feature based optical mapping ［J］. *International Journal of Intelligent Robotics and Applications*, 2019, 3 (2): 196 – 206.

［33］ DACUNTO M, PIERI G, RIGHI M, *et al*. A methodological approach for combining super resolution and pattern recognition to image identification ［J］. *Pattern Recognition and Image Analysis*, 2014, 24 (2): 209 – 217.

［34］ FATANIYA B, ZAVERI T, ACHARYA S. Classification of microscopic image of herbal plants from its powder using speeded-up robust features ［J］. *Journal of Advanced Microscopy Research*, 2018, 13 (3): 326 – 332.

［35］ YIGIT E, DEMIRCI S, UNAL A, *et al*. Millimeter wave ground based synthetic aperture radar imaging for foreign object debris detection: experimental studies at short ranges ［J］. *Journal of Infrared, Millimeter and Terahertz Waves*, 2012, 33 (12): 1227 – 1238.

［36］ XU F, WANG H, HU B, *et al*. Road boundaries detection based on modified occupancy grid map using millimeter wave radar ［J］. *Mobile Networks and Applications*, 2019 (10): 1 – 8.

［37］ WANG J, GE J, ZHANG Q, *et al*. Study of aircraft icing warning algorithm based on millimeter wave radar ［J］. *Journal of Meteorological Research*, 2017, 31 (6): 1034 – 1044.

［38］ YIGIT E. Compressed sensing for millimeter wave ground based SAR/ISAR imaging ［J］. *Journal of Infrared, Millimeter and Terahertz Waves*, 2014, 35 (11): 932 – 948.

［39］ 王宝帅, 兰竹, 李正杰, 等. 毫米波雷达机场跑道异物分层检测算法 ［J］. 电子与信息学报, 2018, 40 (11): 2676 – 2683.

［40］ 吴晓彪, 孙合敏, 吴卫华, 等. 一种基于车载 SAR 的 FOD 检测系统及成像性能分析 ［J］. 空军预警学院学报, 2018, 32 (4): 263 – 266.

［41］ XU H, HAN Z, FENG S, *et al*. Foreign object debris material recognition based on convolutional neural networks ［J］. *Journal on Image and Video Processing*, 2018 (12): 21 – 27.

［42］ DING P, ZHANG Y, JIA P, *et al*. A comparison: different DCNN models for intelligent object detection in remote sensing images ［J］. *Neural Processing Letters*, 2019, 49 (3): 1369 – 1379.

［43］ LONG T, LIANG Z, LIU Q. Advanced technology of high resolution radar: target detection, tracking, imaging and recognition ［J］. *Science China, Information Sciences*, 2019, 62: 40301 – 40308.

［44］ LIU K, LIU D. Particle tracking velocimetry and flame front detection techniques on commercial aircraft debris striking events ［J］. *Journal of Visualization*, 2019, 22 (4): 783 – 794.

［45］ RAMIK D M, SABOURIN C, MORENO R, *et al*. A machine learning based intelligent vision system for autonomous object detection and recognition ［J］. *Applied Intelligence*, 2014, 40 (2): 358 – 375.

［46］ 张思睿，葛万成，汪亮友. 恶劣天气下可见光和红外图像融合算法设计 ［J］. 信息技术，2016, 6 (1): 33 – 36.

［47］ 张群，何其芳，罗迎. 基于贝塞尔函数基信号分解的微动群目标特征提取方法 ［J］. 电子与信息学报，2016, 38 (12): 3056 – 3062.

［48］ 李煜，肖刚. 机场跑道异物检测系统设计与研究 ［J］. 激光与红外，2011, 41 (8): 909 – 915.

［49］ 党国龙，孙瑾，杨刘涛，等. 基于改进视觉注意机制的跑道异物检测算法 ［J］. 航空计算技术，2018, 48 (3): 99 – 102.

［50］ YUAN X. Image multitarget detection and segmentation algorithm based on regional proposed fast intelligent network ［J］. *Cluster Computing*, 2019, 22 (s2): 3385 – 3393.

［51］ 兰红，韩纪东，方毅. 活动轮廓模型在医学图像分割的综述 ［J］. 科学技术与工程，2018, 18 (16): 161 – 167.

［52］ 李倩，江泽涛. 二值化的 SIFT 特征描述子及图像拼接优化 ［J］. 中国图象图形学报，2016, 21 (12): 1593 – 1601.

［53］ 唐涛，覃晓，易宗剑，等. 基于 K 中心点聚类的图像二值化方法 ［J］. 计算机科学与探索，2015 (2): 234 – 241.

［54］ 杨陶，田怀文，刘晓敏，等. 基于边缘检测与 OTSU 的图像分割算法研究 ［J］. 计算机工程，2016, 42 (11): 255 – 260.

［55］ LIU G G, NIU F J, WU Z W. Life cycle performance prediction for rigid runway pavement using artificial neural network ［J］. *International Journal of Pavement Engineering*, 2020, 21 (14): 1806 – 1814.

［56］ YE W, YU Y J. Greedy algorithm for the design of linear phase fir filters with sparse coefficients ［J］. *Circuits, Systems and Signal Processing*, 2016, 35 (4): 1427 – 1436.

［57］ 赵宇明，熊蕙霖，周越，等. 模式识别 ［M］. 上海：上海交通大学出版社，2013.

［58］ THEODORIDIS S, KOUTROUMBAS K. 模式识别 ［M］. 李晶皎，王爱侠，王骄，等，译. 4 版. 北京：电子工业出版社，2020.

［59］ 陈志仁，顾红，苏卫民，等. 低分辨雷达目标分类的最小代价拒判算法 ［J］. 自

动化学报，2018，44（6）：1062 –1071.

[60] MERKULOV V I, SADOVSKII P A. A measurement identification algorithm for a multiple object tracking system in multiband radar systems ［J］. *Journal of Communications Technology and Electronics*, 2016, 61（10）：1101 –1106.

[61] MKRTCHYANG A. Determination of the scattering matrix of multiple target by the received OFDM radar signal ［J］. *Journal of Contemporary Physics*, 2019, 54（1）：77 –83.

[62] 冯德军，谢前朋，王俊杰，等. 对雷达回波的无源电磁调控技术及其发展 ［J］. 系统工程与电子技术，2019，41（6）：1236 –1241.

[63] HE F Z, LI X. Research on occluded face recognition based on low rank sparse and network learning ［J］. *Computer Simulation*, 2020, 37（10）：14 –17.

[64] YANG R, DUAN Z, LU Y, *et al*. Load reduction test method of similarity theory and BP neural networks of large cranes ［J］. *Chinese Journal of Mechanical Engineering*, 2016, 29（1）：145 –151.

[65] ZHANG Q, ZHU S. Visual interpretability for deep learning: a survey ［J］. *Frontiers of Information Technology & Electronic Engineering*, 2018, 19（1）：27 –39.

[66] ZHOU Y, GAO Z. Intelligent recognition of medical motion image combining convolutional neural network with internet of things ［J］. *IEEE Access*, 2019, 7（99）：462 –476.

[67] 林景栋，吴欣怡，柴毅，等. 卷积神经网络结构优化综述 ［J］. 自动化学报，2020，46（1）：24 –37.

[68] SHAN S, HU Z, LIU, Z, *et al*. An adaptive genetic algorithm for demand driven and resource constrained project scheduling in aircraft assembly ［J］. *Information Technology and Management*, 2017, 18（1）：41 –53.

[69] HERNANDEZ L P, KARTAL B, TAYLOR M E. A survey and critique of multiagent deep reinforcement learning ［J］. *Autonomous Agents and Multi-Agent Systems*, 2019, 33（6）：750 –797.

[70] WANG H, WANG T, ZHOU Y, *et al*. Information classification algorithm based on decision tree optimization ［J］. *Cluster Computing*, 2019, 22（s3）：7559 –7568.

[71] GOODFELLOW I, BENGIO Y, COURVILLE A. 深度学习 ［M］. 赵申剑，黎或君，符天凡，等，译. 北京：人民邮电出版社，2017.

[72] ZHENG Y P, LI G Y, LI Y. A review on the application of deep learning in image recognition ［J］. *Computer Engineering and Applications*, 2019, 55（12）：20 –36.

[73] DING S, ZHAO X, XU X, *et al*. An effective asynchronous framework for small scale reinforcement learning problems ［J］. *Applied Intelligence*, 2019, 49（12）：4303 –4318.

[74] ANDERSON S J, KRIGOLSON O E, JAMNICZKY H A, *et al*. Learning anatomical structures: a reinforcement based learning approach ［J］. *Medical Science Educator*,

2016, 26（1）：123 – 128.

［75］ 王涛，张化光. 基于策略迭代的连续时间系统的随机线性二次最优控制 ［J］. 控制与决策，2015，30（9）：1674 – 1678.

［76］ PAN W, QU R, HWANG K S, et al. An ensemble fuzzy approach for inverse reinforcement learning ［J］. International Journal of Fuzzy Systems, 2019, 21（1）：95 – 103.

［77］ KANGASRAASIO A, KASKI S. Inverse reinforcement learning from summary data ［J］. Machine Learning, 2018, 107（8 – 10）：1517 – 1535.

［78］ 陈思，孙有朝，郑敏. 基于支持向量机的飞机重着陆风险预警模型 ［J］. 兵器装备工程学报，2019，40（9）：154 – 158.

［79］ KHODKAR Z, ALAVI S M. Target classification enhancement in vhf radar using support vector machine ［J］. Iranian Journal of Science and Technology, Transactions of Electrical Engineering, 2016, 40（1）：51 – 62.

［80］ WU Y, FENG J. Development and application of artificial neural network ［J］. Wireless Personal Communications, 2018, 102：1645 – 1656.

［81］ MODARESI F, ARAGHINEJAD S, EBRAHIMI K. A comparative assessment of artificial neural network, generalized regression neural network, least square support vector regression and k – nearest neighbor regression for monthly streamflow forecasting in linear and nonlinear conditions ［J］. Water Resources Management, 2018, 32：243 – 258.

［82］ EGORCHEV M V, TIUMENTSEV Y V. Learning of semi-empirical neural network model of aircraft three axis rotational motion ［J］. Optical Memory and Neural Networks, 2015, 24（3）：201 – 208.

［83］ SOUZA J P C, MARCATO A L M, AGUIAR E P, et al. Autonomous landing of UAV based on artificial neural network supervised by fuzzy logic ［J］. Journal of Control, Automation and Electrical Systems, 2019, 30（4）：522 – 531.

［84］ ENGELEN J E, HOOS H H. A survey on semi-supervised learning ［J］. Machine Learning, 2019（11）：1 – 68.

［85］ BAGHERZADEH J, ASIL H. A review of various semi-supervised learning models with a deep learning and memory approach ［J］. Iran Journal of Computer Science, 2019, 2（2）：65 – 80.

［86］ CAO L, TANG S, ZHANG D. Flight control for air breathing hypersonic vehicles using linear quadratic regulator design based on stochastic robustness analysis ［J］. Frontiers of Information Technology & Electronic Engineering, 2017, 18（7）：882 – 897.

［87］ WANG Z J, TURKO R, SHAIKH O, et al. CNN explainer：learning convolutional neural networks with interactive visualization ［J］. IEEE Transactions on Visualization and Computer Graphics, 2021, 27：1396 – 1406.

［88］ HU G, YANG Z, HAN J, et al. Aircraft detection in remote sensing images based on saliency and convolution neural network ［J］. Journal on Wireless Communications and

Networking，2018（12）：26–30.

［89］ REGMI H K, NESAMONY J, PAPPADA S M, *et al*. A system for realtime syringe classification and volume measurement using a combination of image processing and artificial neural networks ［J］. *Journal of Pharmaceutical Innovation*, 2019, 14 (4): 341–358.

［90］ 左兴，刘洪兵. 改进神经网络对机场刚性道面损伤参数的识别 ［J］. 计算机仿真，2012，29（8）：50–54.

［91］ 马洪超，郭丽艳. 人工神经网络信息融合及其在机场识别中的应用研究 ［J］. 武汉大学学报（信息科学版），2005，30（8）：682–684.

［92］ YILDIRIM M, KACAR F. Adapting laplacian based filtering in digital image processing to a retina inspired analog image processing circuit ［J］. *Analog Integrated Circuits and Signal Processing*, 2019, 100 (3): 537–545.

［93］ MAKTAV D, BERBEROGLU S. Different digital image processing methods for remote sensing applications ［J］. *Journal of the Indian Society of Remote Sensing*, 2018, 46 (8): 1201–1202.

［94］ 程腾，蒋亚西，吴勃夫，等. 一种基于数字图像相关的高精度车辆跟踪算法 ［J］. 汽车工程，2018，40（8）：942–946.

［95］ 章毓晋. 图像工程 ［M］. 北京：清华大学出版社，2018.

［96］ 孙挺，齐迎春，耿国华. 基于帧间差分和背景差分的运动目标检测算法 ［J］. 吉林大学学报（工学版），2016，46（4）：1325–1329.

［97］ 侯春萍，张倩文，王晓燕，等. 轮廓匹配的复杂背景中目标检测算法 ［J］. 哈尔滨工业大学学报，2020，52（5）：121–128.

［98］ 朱婧. 基于改进的光流场算法对运动目标的检测与跟踪技术研究 ［D］. 哈尔滨：哈尔滨工程大学，2006.

［99］ 张西顺. 图像的亚像素分割技术研究及在飞机外轮廓分割中的应用 ［D］. 北京：北京交通大学，2019.

［100］ 杜兴强. 基于深度背景差分的铁路异物检测算法 ［D］. 北京：北京交通大学，2019.

［101］ 贺东风，赵越让，郭博智，等. 中国商用飞机有限责任公司系统工程手册 ［M］. 上海：上海交通大学出版社，2019.

［102］ 陈迎春，宋文滨，刘洪. 民用飞机总体设计 ［M］. 上海：上海交通大学出版社，2019.

［103］ GLAHN C, KLUIN M, HERMANN I, *et al*. Requirements for the boosting system of future engine concepts ［J］. *MTZ Worldwide*, 2017, 78 (4): 16–21.

［104］ ZHANG D, WEI B. Convergence performance comparisons of PID, MRAC and PID MRAC hybrid controller ［J］. *Frontiers of Mechanical Engineering*, 2016, 11 (2): 213–217.

［105］ MARJANI A, SHIRAZIAN S, ASADOLLAHZADEH M. Topology optimization of neural networks based on a coupled genetic algorithm and particle swarm optimization techniques [J]. *Neural Computing and Applications*, 2018, 29 (11): 1073 – 1076.

[106] MOBAYEN S. An adaptive fast terminal sliding mode control combined with global sliding mode scheme for tracking control of uncertain nonlinear third order systems [J]. *Nonlinear Dynamics*, 2015, 82 (1 – 2): 599 – 610.

[107] LIU C L, WANG S T, KUO S C. An intelligent fuzzy control system with adapted interval for improving the supervisory performance in automation [J]. *Operational Research*, 2018, 18 (3): 689 – 709.

[108] PIROOZ M, FATEH M M. Impedance fuzzy control of an active aircraft landing gear system [J]. *International Journal of Dynamics and Control*, 2019, 7 (4): 1392 – 1403.

[109] LI T, MA S, LI B, *et al*. Fuzzy theory based control method for an in-pipe robot to move in variable resistance environment [J]. *Chinese Journal of Mechanical Engineering*, 2015, 28 (6): 1213 – 1221.

[110] ARIK S O, PFISTER T. Proto attend: attention based prototypical learning [J]. *Journal of Machine Learning Research*, 2020, 21 (210): 1 – 3.

[111] ARUNKUMAR P, SHANTHARAJAH S P, GEETHA M. Improved canny detection algorithm for processing and segmenting text from the images [J]. *Cluster Computing*, 2019, 22 (s3): 7015 – 7021.

[112] 仲伟波，姚旭洋，冯友兵，等. 双目区域视差快速计算及测距算法 [J]. 中国图象图形学报, 2019, 24 (9): 1537 – 1545.

[113] 吴骏，李文杰，耿磊，等. 基于单目视觉的前方车辆检测与测距 [J]. 计算机工程, 2017, 43 (2): 26 – 32.

[114] MANORANJITHAM R, DEEPA P. Efficient invariant interest point detector using bilateral Harris corner detector for object recognition application [J]. *Multimedia Tools and Applications*, 2018, 77 (8): 9365 – 9378.

[115] ZHENG F, LUO S, SONG K, *et al*. Improved lane line detection algorithm based on hough transform [J]. *Pattern Recognition and Image Analysis*, 2018, 28 (2): 254 – 260.

[116] XU K, LIU J, MIAO J, *et al*. An improved SIFT algorithm based on adaptive fractional differential [J]. *Journal of Ambient Intelligence and Humanized Computing*, 2019, 10 (8): 3297 – 3305.

[117] ALI H S, ISMAIL A I, FARAG F A, *et al*. Speeded up robust features for efficient iris recognition [J]. *Signal, Image and Video Processing*, 2016, 10 (8): 1385 – 1391.

[118] 刘颖，雷研博，范九伦，等. 基于小样本学习的图像分类技术综述 [J]. 自动化学报, 2021, 47 (2): 297 – 315.

[119] 邓雪峰，孙瑞志，张永瀚，等. 基于数据位图的滑动分块算法 [J]. 计算机研究与发展, 2014, 51 (s1): 30 – 38.

[120] CHANDRA S, MAHESHKAR S. Verification of static signature pattern based on random subspace, REP tree and bagging [J]. *Multimedia Tools and Applications*, 2017, 76

(18): 19139 – 19171.

[121] SERBAN A, POLL E, VISSERJ. Adversarial examples on object recognition: a comprehensive survey [J]. *ACM Computing Surveys*, 2020, 53 (3): 1 – 39.

[122] SREEKUMARI P, JUNG J. Transport protocols for data center networks: a survey of issues, solutions and challenges [J]. *Photonic Network Communications*, 2016, 31 (1): 112 – 128.

[123] THOMBRE S K, PATNAIK L M, TAVILDAR A S. Effect of multiplexing TCP flows on delay sensitivity in the internet communications [J]. *CCF Transactions on Networking*, 2019, 1 (1 – 4): 37 – 51.

[124] ZHANG Y, CHEN X. Explainable recommendation: a survey and new perspectives [J]. *Foundations and Trends in Information Retrieval*, 2020, 14 (1): 1 – 101.

[125] ARRIETA A B, DIAZ R N, SER J D, *et al*. Explainable artificial intelligence (XAI): concepts, taxonomies, opportunities and challenges toward responsible AI [J]. *Information Fusion*, 2020, 58: 82 – 115.

[126] XU J, TIAN W J, FAN Y Y. Simulation of face key point recognition and location method based on deep learning [J]. *Computer Simulation*, 2020, 37 (6): 434 – 438.